Physical property data book

For engineers and scientists

David Shallcross
Department of Chemical and Biomolecular Engineering
University of Melbourne, Australia

For Peter, Emily and Andrew

IChem**E**
INSTITUTION OF CHEMICAL ENGINEERS

Published by
Institution of Chemical Engineers (IChemE)
Davis Building
165–189 Railway Terrace
Rugby
Warwickshire CV21 3HQ, UK

IChemE is a Registered Charity
Offices in Rugby (UK), London (UK) and Melbourne (Australia)

© 2004 David Shallcross

ISBN 0 85295 465 4

Typeset by Techset Composition Limited, Salisbury, UK
Printed by Hobbs the Printers Limited, Totton, Hampshire, UK

ii

Preface

As an undergraduate engineering student in the late 1970s, one of the best purchases I made was a slim book of steam tables and other physical property data. This book followed me around the world, always being close to hand. As an academic I have always strongly encouraged my students to purchase a set of steam tables and have developed my lecturing material on the assumption that the students have ready access to such valuable resources. In more recent years I have come to feel that the existing tables lack much useful data, and that what data are present may be very out of date. I therefore decided to develop my own set of tables that would be of use to students and practising engineers and scientists alike. This book is the result.

The book contains the latest information on the properties of water and steam, air and six different refrigerants. Where the reader may need access to more comprehensive data for a particular material, the appropriate references are supplied. Some of the more useful diagrams that students, engineers and scientists require are provided in a clear format. In some cases this information is presented in SI units for the first time anywhere.

I have used a form of this book for the last three years in my teaching and have found it a useful resource. I trust that others will value it and that students buying it today will still find themselves reaching for it in 30 years' time.

David Shallcross
September 2004

Contents

Periodic table of the elements

Legend: Atomic number → 28 **Ni** ← Symbol · Name → 58.6934 **Nickel** → Atomic weight

1 H 1.00794 Hydrogen																	2 He 4.00260 Helium
3 Li 6.941 Lithium	4 Be 9.01218 Beryllium											5 B 10.811 Boron	6 C 12.0107 Carbon	7 N 14.0067 Nitrogen	8 O 15.9994 Oxygen	9 F 18.9984 Fluorine	10 Ne 20.1797 Neon
11 Na 22.9898 Sodium	12 Mg 24.3050 Magnesium											13 Al 26.9815 Aluminium	14 Si 28.0855 Silicon	15 P 30.9738 Phosphorus	16 S 32.065 Sulfur	17 Cl 35.453 Chlorine	18 Ar 39.948 Argon
19 K 39.0983 Potassium	20 Ca 40.078 Calcium	21 Sc 44.9559 Scandium	22 Ti 47.867 Titanium	23 V 50.9415 Vanadium	24 Cr 51.9961 Chromium	25 Mn 54.9380 Manganese	26 Fe 55.845 Iron	27 Co 58.9332 Cobalt	28 Ni 58.6934 Nickel	29 Cu 63.546 Copper	30 Zn 65.39 Zinc	31 Ga 69.723 Gallium	32 Ge 72.64 Germanium	33 As 74.9216 Arsenic	34 Se 78.96 Selenium	35 Br 79.904 Bromine	36 Kr 83.80 Krypton
37 Rb 85.4678 Rubidium	38 Sr 87.62 Strontium	39 Y 88.9059 Yttrium	40 Zr 91.224 Zirconium	41 Nb 92.9064 Niobium	42 Mo 95.94 Molybdenum	43 Tc Technetium	44 Ru 101.07 Ruthenium	45 Rh 102.906 Rhodium	46 Pd 106.42 Palladium	47 Ag 107.868 Silver	48 Cd 112.411 Cadmium	49 In 114.818 Indium	50 Sn 118.710 Tin	51 Sb 121.760 Antimony	52 Te 127.60 Tellurium	53 I 126.904 Iodine	54 Xe 131.29 Xenon
55 Cs 132.905 Cesium	56 Ba 137.327 Barium	57 La 138.906 Lanthanum	72 Hf 178.49 Hafnium	73 Ta 180.948 Tantalum	74 W 183.84 Tungsten	75 Re 186.207 Rhenium	76 Os 190.23 Osmium	77 Ir 192.217 Iridium	78 Pt 195.078 Platinum	79 Au 196.967 Gold	80 Hg 200.59 Mercury	81 Tl 204.383 Thallium	82 Pb 207.2 Lead	83 Bi 208.980 Bismuth	84 Po Polonium	85 At Astatine	86 Rn Radon
87 Fr Francium	88 Ra Radium	89 Ac Actinium	104 Rf Rutherfordium	105 Ha Hahnium													

Lanthanides

58 Ce 140.116 Cerium	59 Pr 140.908 Praseodymium	60 Nd 144.24 Neodymium	61 Pm Promethium	62 Sm 150.36 Samarium	63 Eu 151.964 Europium	64 Gd 157.25 Gadolinium	65 Tb 158.925 Terbium	66 Dy 162.50 Dysprosium	67 Ho 164.930 Holmium	68 Er 167.26 Erbium	69 Tm 168.934 Thulium	70 Yb 173.04 Ytterbium	71 Lu 174.967 Lutetium

Actinides

90 Th 232.038 Thorium	91 Pa 231.036 Protactinium	92 U 238.029 Uranium	93 Np Neptunium	94 Pu Plutonium	95 Am Americium	96 Cm Curium	97 Bk Berkelium	98 Cf Californium	99 Es Einsteinium	100 Fm Fermium	101 Md Mendelevium	102 No Nobelium	103 Lr Lawrencium

Atomic weights taken from Coplen, T.B., 2001, *J. Phys. Chem. Ref. Data*, **30(3)**: 701–712.

Saturated water and steam properties (temperature)

t (°C)	P (kPa)	v_g (m³ kg⁻¹)	h_f (kJ kg⁻¹)	h_g (kJ kg⁻¹)	h_{fg} (kJ kg⁻¹)	S_f (kJ kg⁻¹ K⁻¹)	S_g (kJ kg⁻¹ K⁻¹)	S_{fg} (kJ kg⁻¹ K⁻¹)
0.01	0.6117	205.991	0.00	2500.9	2500.9	0.0000	9.155	9.155
1	0.6571	192.439	4.18	2502.7	2498.6	0.0153	9.129	9.114
2	0.7060	179.758	8.39	2504.6	2496.2	0.0306	9.103	9.072
3	0.7581	168.008	12.60	2506.4	2493.8	0.0459	9.076	9.031
4	0.8135	157.116	16.81	2508.2	2491.4	0.0611	9.051	8.989
5	0.8726	147.011	21.02	2510.1	2489.0	0.0763	9.025	8.949
6	0.9354	137.633	25.22	2511.9	2486.7	0.0913	8.999	8.908
7	1.0021	128.923	29.43	2513.7	2484.3	0.1064	8.974	8.868
8	1.0730	120.829	33.63	2515.6	2481.9	0.1213	8.949	8.828
9	1.1483	113.304	37.82	2517.4	2479.6	0.1362	8.924	8.788
10	1.2282	106.303	42.02	2519.2	2477.2	0.1511	8.900	8.749
11	1.3130	99.787	46.22	2521.0	2474.8	0.1659	8.875	8.710
12	1.4028	93.719	50.41	2522.9	2472.5	0.1806	8.851	8.671
13	1.4981	88.064	54.60	2524.7	2470.1	0.1953	8.827	8.632
14	1.5990	82.793	58.79	2526.5	2467.7	0.2099	8.804	8.594
15	1.7058	77.876	62.98	2528.3	2465.4	0.2245	8.780	8.556
16	1.8188	73.286	67.17	2530.2	2463.0	0.2390	8.757	8.518
17	1.9384	69.001	71.36	2532.0	2460.6	0.2534	8.734	8.481
18	2.0647	64.998	75.54	2533.8	2458.3	0.2678	8.711	8.443
19	2.1983	61.256	79.73	2535.6	2455.9	0.2822	8.688	8.406
20	2.3393	57.757	83.91	2537.4	2453.5	0.2965	8.666	8.370
21	2.4882	54.483	88.10	2539.3	2451.2	0.3107	8.644	8.333
22	2.6453	51.418	92.28	2541.1	2448.8	0.3249	8.622	8.297
23	2.8111	48.548	96.46	2542.9	2446.4	0.3391	8.600	8.261
24	2.9858	45.858	100.65	2544.7	2444.0	0.3532	8.578	8.225
25	3.1699	43.337	104.83	2546.5	2441.7	0.3672	8.557	8.189
26	3.3639	40.973	109.01	2548.3	2439.3	0.3812	8.535	8.154
27	3.5681	38.754	113.19	2550.1	2436.9	0.3952	8.514	8.119
28	3.7831	36.672	117.37	2551.9	2434.6	0.4091	8.493	8.084
29	4.0092	34.716	121.55	2553.7	2432.2	0.4229	8.473	8.050
30	4.2470	32.878	125.73	2555.5	2429.8	0.4368	8.452	8.015
32	4.7596	29.527	134.09	2559.2	2425.1	0.4642	8.411	7.947
34	5.3251	26.560	142.45	2562.8	2420.3	0.4915	8.371	7.880
36	5.9479	23.929	150.81	2566.3	2415.5	0.5187	8.332	7.813
38	6.6328	21.593	159.17	2569.9	2410.8	0.5456	8.294	7.748
40	7.3849	19.515	167.53	2573.5	2406.0	0.5724	8.256	7.683
42	8.2096	17.664	175.89	2577.1	2401.2	0.5990	8.218	7.619
44	9.1124	16.011	184.25	2580.6	2396.4	0.6255	8.181	7.556
46	10.0994	14.534	192.62	2584.2	2391.6	0.6517	8.145	7.494
48	11.1771	13.212	200.98	2587.8	2386.8	0.6779	8.110	7.432
50	12.352	12.027	209.34	2591.3	2381.9	0.7038	8.075	7.371
52	13.631	10.963	217.71	2594.8	2377.1	0.7296	8.040	7.311
54	15.022	10.006	226.07	2598.3	2372.3	0.7553	8.007	7.251
56	16.533	9.1448	234.44	2601.8	2367.4	0.7808	7.973	7.192
58	18.171	8.3683	242.81	2605.3	2362.5	0.8061	7.940	7.134
60	19.946	7.6672	251.18	2608.8	2357.7	0.8313	7.908	7.077
65	25.042	6.1935	272.12	2617.5	2345.4	0.8937	7.830	6.936
70	31.201	5.0395	293.07	2626.1	2333.0	0.9551	7.754	6.799
75	38.595	4.1289	314.03	2634.6	2320.6	1.0158	7.681	6.665

Saturated water and steam properties (temperature) (cont.)

t (°C)	P (kPa)	v_g (m³ kg⁻¹)	h_f (kJ kg⁻¹)	h_g (kJ kg⁻¹)	h_{fg} (kJ kg⁻¹)	S_f (kJ kg⁻¹ K⁻¹)	S_g (kJ kg⁻¹ K⁻¹)	S_{fg} (kJ kg⁻¹ K⁻¹)
80	47.414	3.4052	335.01	2643.0	2308.0	1.0756	7.611	6.535
85	57.867	2.8258	356.01	2651.3	2295.3	1.1346	7.543	6.409
90	70.182	2.3591	377.04	2659.5	2282.5	1.1929	7.478	6.285
95	84.608	1.9806	398.09	2667.6	2269.5	1.2504	7.415	6.165
100*	101.418	1.6718	419.17	2675.6	2256.4	1.3072	7.354	6.047
105	120.90	1.4184	440.3	2683.4	2243.1	1.363	7.295	5.932
110	143.38	1.2093	461.4	2691.1	2229.6	1.419	7.238	5.819
115	169.18	1.0358	482.6	2698.6	2216.0	1.474	7.183	5.709
120	198.67	0.89121	503.8	2705.9	2202.1	1.528	7.129	5.601
125	232.24	0.77003	525.1	2713.1	2188.0	1.582	7.077	5.495
130	270.28	0.66800	546.4	2720.1	2173.7	1.635	7.026	5.392
135	313.23	0.58173	567.7	2726.9	2159.1	1.687	6.977	5.290
140	361.54	0.50845	589.2	2733.4	2144.3	1.739	6.929	5.190
145	415.69	0.44596	610.6	2739.8	2129.2	1.791	6.883	5.092
150	476.17	0.39245	632.2	2745.9	2113.7	1.842	6.837	4.995
155	543.50	0.34646	653.8	2751.8	2098.0	1.892	6.793	4.900
160	618.24	0.30678	675.5	2757.4	2082.0	1.943	6.749	4.807
165	700.93	0.27243	697.2	2762.8	2065.6	1.992	6.707	4.714
170	792.19	0.24259	719.1	2767.9	2048.8	2.042	6.665	4.623
175	892.60	0.21658	741.0	2772.7	2031.7	2.091	6.624	4.533
180	1002.8	0.19384	763.1	2777.2	2014.2	2.139	6.584	4.445
185	1123.5	0.17390	785.2	2781.4	1996.2	2.188	6.545	4.357
190	1255.2	0.15636	807.4	2785.3	1977.9	2.236	6.506	4.270
195	1398.8	0.14089	829.8	2788.8	1959.0	2.283	6.468	4.185
200	1554.9	0.12721	852.3	2792.0	1939.7	2.331	6.430	4.100
210	1907.7	0.10429	897.6	2797.3	1899.6	2.425	6.356	3.932
220	2319.6	0.086092	943.6	2800.9	1857.4	2.518	6.284	3.766
230	2797.1	0.071504	990.2	2802.9	1812.7	2.610	6.213	3.603
240	3346.9	0.059705	1037.6	2803.0	1765.4	2.702	6.142	3.440
250	3976.2	0.050083	1085.8	2800.9	1715.2	2.794	6.072	3.279
260	4692.3	0.042173	1135.0	2796.6	1661.6	2.885	6.002	3.117
270	5503.0	0.035621	1185.3	2789.7	1604.4	2.977	5.930	2.954
280	6416.6	0.030153	1236.9	2779.9	1543.0	3.069	5.858	2.789
290	7441.8	0.025555	1290.0	2766.7	1476.7	3.161	5.783	2.622
300	8587.9	0.021660	1345.0	2749.6	1404.6	3.255	5.706	2.451
310	9865.1	0.018335	1402.2	2727.9	1325.7	3.351	5.624	2.273
320	11,284	0.015471	1462.2	2700.6	1238.4	3.449	5.537	2.088
330	12,858	0.012979	1525.9	2666.0	1140.2	3.552	5.442	1.890
340	14,601	0.010781	1594.5	2621.8	1027.3	3.660	5.336	1.675
350	16,529	0.008802	1670.9	2563.6	892.7	3.778	5.211	1.433
360	18,666	0.006949	1761.7	2481.5	719.8	3.917	5.054	1.137
365	19,821	0.006012	1817.8	2422.9	605.2	4.001	4.950	0.948
370	21,044	0.004954	1890.7	2334.5	443.8	4.111	4.801	0.690
373.95	22,064	0.00311	2084.3	2084.3	0.0	4.407	4.407	0.000

Enthalpy datum condition is triple point of ice, water and steam, 0.01°C and 0.612 kPa.
*The vapour pressure of water vapour at 100°C is 101.325 kPa. The equation upon which the tabulated data is based (see page 24) calculates a vapour pressure of 101.418 kPa, an error of less than 0.1%. See entry on page 4 for data at vapour pressure of 101.325 kPa.

Saturated water and steam properties (pressure)

P (kPa)	t (°C)	v_g (m³ kg⁻¹)	h_f (kJ kg⁻¹)	h_g (kJ kg⁻¹)	h_{fg} (kJ kg⁻¹)	S_f (kJ kg⁻¹ K⁻¹)	S_g (kJ kg⁻¹ K⁻¹)	S_{fg} (kJ kg⁻¹ K⁻¹)
0.6117	0.1	205.991	0.00	2500.9	2500.9	0.0000	9.155	9.155
1.0	7.0	129.178	29.30	2513.7	2484.4	0.1059	8.975	8.869
1.5	13.0	87.959	54.68	2524.7	2470.0	0.1956	8.827	8.631
2.0	17.5	66.987	73.43	2532.9	2459.4	0.2606	8.723	8.462
2.5	21.1	54.240	88.42	2539.4	2451.0	0.3118	8.642	8.330
3.0	24.1	45.653	100.98	2544.8	2443.9	0.3543	8.576	8.222
3.5	26.7	39.466	111.82	2549.5	2437.7	0.3906	8.521	8.131
4.0	29.0	34.791	121.39	2553.7	2432.3	0.4224	8.473	8.051
4.5	31.0	31.131	129.96	2557.4	2427.4	0.4507	8.431	7.981
5.0	32.9	28.185	137.75	2560.7	2423.0	0.4762	8.394	7.918
6.0	36.2	23.733	151.48	2566.6	2415.2	0.5208	8.329	7.808
7.0	39.0	20.524	163.35	2571.7	2408.4	0.5590	8.274	7.715
8.0	41.5	18.099	173.84	2576.2	2402.4	0.5925	8.227	7.635
9.0	43.8	16.199	183.25	2580.2	2397.0	0.6223	8.186	7.564
10.0	45.8	14.670	191.81	2583.9	2392.1	0.6492	8.149	7.500
12	49.4	12.358	206.91	2590.3	2383.4	0.6963	8.085	7.389
14	52.5	10.691	219.99	2595.8	2375.8	0.7366	8.031	7.294
16	55.3	9.4306	231.57	2600.6	2369.1	0.7720	7.985	7.213
18	57.8	8.4431	241.96	2605.0	2363.0	0.8035	7.944	7.140
20	60.1	7.6480	251.42	2608.9	2357.5	0.8320	7.907	7.075
22	62.1	6.9936	260.11	2612.5	2352.4	0.8580	7.874	7.016
24	64.1	6.4453	268.15	2615.9	2347.7	0.8819	7.844	6.962
26	65.8	5.9792	275.64	2619.0	2343.3	0.9041	7.817	6.913
28	67.5	5.5778	282.66	2621.8	2339.2	0.9247	7.791	6.866
30	69.1	5.2284	289.27	2624.5	2335.3	0.9441	7.767	6.823
32	70.6	4.9215	295.52	2627.1	2331.6	0.9623	7.745	6.783
34	72.0	4.6497	301.45	2629.5	2328.1	0.9795	7.725	6.745
36	73.3	4.4072	307.09	2631.8	2324.7	0.9958	7.705	6.709
38	74.6	4.1895	312.47	2634.0	2321.5	1.0113	7.687	6.675
40	75.9	3.9930	317.62	2636.1	2318.4	1.0261	7.669	6.643
50	81.3	3.2400	340.54	2645.2	2304.7	1.0912	7.593	6.502
60	85.9	2.7317	359.91	2652.9	2292.9	1.1454	7.531	6.386
70	89.9	2.3648	376.75	2659.4	2282.7	1.1921	7.479	6.287
80	93.5	2.0871	391.71	2665.2	2273.5	1.2330	7.434	6.201
90	96.7	1.8694	405.20	2670.3	2265.1	1.2696	7.394	6.125
100	99.6	1.6939	417.50	2674.9	2257.4	1.3028	7.359	6.056
101.325	100.0	1.6732	419.06	2675.5	2256.5	1.3069	7.354	6.048
110	102.3	1.5495	428.8	2679.2	2250.3	1.3330	7.327	5.994
120	104.8	1.4284	439.4	2683.1	2243.7	1.3609	7.298	5.937
130	107.1	1.3253	449.2	2686.6	2237.5	1.3868	7.271	5.884
140	109.3	1.2366	458.4	2690.0	2231.6	1.4110	7.246	5.835
150	111.3	1.1593	467.1	2693.1	2226.0	1.4337	7.223	5.789
160	113.3	1.0914	475.4	2696.0	2220.7	1.4551	7.201	5.746
170	115.1	1.0312	483.2	2698.8	2215.6	1.4753	7.181	5.706
180	116.9	0.97747	490.7	2701.4	2210.7	1.4945	7.162	5.668
190	118.6	0.92924	497.9	2703.9	2206.0	1.5127	7.144	5.631
200	120.2	0.88568	504.7	2706.2	2201.5	1.5302	7.127	5.597
220	123.2	0.81007	517.6	2710.6	2193.0	1.5628	7.095	5.532
240	126.1	0.74668	529.6	2714.6	2185.0	1.5930	7.066	5.473
260	128.7	0.69273	540.9	2718.3	2177.4	1.6210	7.039	5.418

4

Saturated water and steam properties (pressure) (cont.)

P (kPa)	t (°C)	v_g (m³ kg⁻¹)	h_f (kJ kg⁻¹)	h_g (kJ kg⁻¹)	h_{fg} (kJ kg⁻¹)	S_f (kJ kg⁻¹ K⁻¹)	S_g (kJ kg⁻¹ K⁻¹)	S_{fg} (kJ kg⁻¹ K⁻¹)
280	131.2	0.64624	551.4	2721.7	2170.3	1.647	7.015	5.367
300	133.5	0.60576	561.4	2724.9	2163.5	1.672	6.992	5.320
320	135.7	0.57017	570.9	2727.8	2157.0	1.695	6.970	5.275
340	137.8	0.53864	579.9	2730.6	2150.7	1.717	6.950	5.233
360	139.8	0.51050	588.5	2733.2	2144.7	1.738	6.931	5.193
380	141.8	0.48522	596.8	2735.7	2139.0	1.758	6.913	5.155
400	143.6	0.46238	604.7	2738.1	2133.4	1.776	6.895	5.119
420	145.4	0.44165	612.3	2740.3	2128.0	1.795	6.879	5.085
440	147.1	0.42274	619.6	2742.4	2122.8	1.812	6.864	5.052
460	148.7	0.40542	626.6	2744.4	2117.7	1.829	6.849	5.020
480	150.3	0.38950	633.5	2746.3	2112.8	1.845	6.834	4.990
500	151.8	0.37481	640.1	2748.1	2108.0	1.860	6.821	4.960
600	158.8	0.31558	670.4	2756.1	2085.8	1.931	6.759	4.828
700	164.9	0.27278	697.0	2762.8	2065.8	1.992	6.707	4.715
800	170.4	0.24034	720.9	2768.3	2047.4	2.046	6.662	4.616
900	175.4	0.21489	742.6	2773.0	2030.5	2.094	6.621	4.527
1000	179.9	0.19436	762.5	2777.1	2014.6	2.138	6.585	4.447
1200	188.0	0.16326	798.3	2783.7	1985.4	2.216	6.522	4.306
1400	195.0	0.14078	830.0	2788.8	1958.9	2.284	6.467	4.184
1600	201.4	0.12374	858.5	2792.8	1934.4	2.343	6.420	4.076
1800	207.1	0.11037	884.5	2795.9	1911.4	2.397	6.377	3.980
2000	212.4	0.09959	908.5	2798.3	1889.8	2.447	6.339	3.892
2200	217.2	0.09070	930.9	2800.1	1869.2	2.492	6.304	3.812
2400	221.8	0.08324	951.9	2801.4	1849.6	2.534	6.271	3.737
2600	226.0	0.07690	971.7	2802.3	1830.7	2.574	6.241	3.667
2800	230.1	0.07143	990.5	2802.9	1812.4	2.611	6.212	3.602
3000	233.9	0.06666	1008.3	2803.2	1794.8	2.646	6.186	3.540
3500	242.6	0.05706	1049.8	2802.6	1752.8	2.725	6.124	3.399
4000	250.4	0.04978	1087.5	2800.8	1713.3	2.797	6.070	3.273
4500	257.4	0.04406	1122.2	2797.9	1675.7	2.862	6.020	3.158
5000	263.9	0.03945	1154.6	2794.2	1639.6	2.921	5.974	3.053
6000	275.6	0.03245	1213.9	2784.6	1570.7	3.028	5.890	2.862
7000	285.8	0.02738	1267.7	2772.6	1505.0	3.122	5.815	2.692
8000	295.0	0.02353	1317.3	2758.7	1441.4	3.208	5.745	2.537
9000	303.3	0.02049	1363.9	2742.9	1379.1	3.287	5.679	2.392
10,000	311.0	0.01803	1408.1	2725.5	1317.4	3.361	5.616	2.255
11,000	318.1	0.01599	1450.4	2706.3	1255.9	3.430	5.554	2.124
12,000	324.7	0.01426	1491.5	2685.4	1194.0	3.497	5.494	1.997
13,000	330.9	0.01278	1531.5	2662.7	1131.2	3.561	5.434	1.873
14,000	336.7	0.01149	1571.0	2637.9	1066.9	3.623	5.373	1.750
15,000	342.2	0.01034	1610.2	2610.7	1000.5	3.685	5.311	1.626
16,000	347.4	0.00931	1649.7	2580.8	931.1	3.746	5.246	1.501
17,000	352.3	0.00837	1690.0	2547.5	857.5	3.808	5.179	1.371
18,000	357.0	0.00750	1732.1	2509.8	777.7	3.872	5.106	1.234
19,000	361.5	0.00668	1777.2	2466.0	688.9	3.940	5.026	1.085
20,000	365.7	0.00587	1827.2	2412.3	585.1	4.016	4.931	0.916
21,000	369.8	0.00500	1887.6	2338.6	451.0	4.106	4.808	0.701
22,064	373.95	0.00311	2084.3	2084.3	0.0	4.407	4.407	0.000

Enthalpy datum condition is triple point of ice, water and steam, 0.01°C and 0.612 kPa.

Superheated steam properties

	T	50	100	150	200	250	300	350
	ρ	0.00671	0.00581	0.00512	0.00458	0.00414	0.00378	0.00348
1.00 kPa	h	2594.4	2688.6	2783.7	2880.0	2977.7	3077.0	3177.7
7.0°C	C_P	1.8761	1.8914	1.9139	1.9403	1.9692	1.9996	2.0312
	S	9.2430	9.5139	9.7531	9.9682	10.165	10.346	10.514
	ρ	0.03358	0.02905	0.02561	0.02290	0.02071	0.01890	0.01739
5.00 kPa	h	2593.3	2688.1	2783.4	2879.8	2977.6	3076.9	3177.6
32.9°C	C_P	1.8964	1.8977	1.9165	1.9417	1.9700	2.0001	2.0315
	S	8.4975	8.7701	9.0098	9.2251	9.4216	9.6027	9.7713
	ρ	0.06726	0.05815	0.05125	0.04582	0.04143	0.03781	0.03478
10.0 kPa	h	2592.0	2687.5	2783.0	2879.6	2977.4	3076.7	3177.5
45.8°C	C_P	1.9280	1.9058	1.9199	1.9434	1.9710	2.0008	2.0319
	S	8.1741	8.4489	8.6892	8.9049	9.1015	9.2827	9.4513
	ρ		0.1165	0.1026	0.09168	0.08289	0.07565	0.06957
20.0 kPa	h		2686.2	2782.3	2879.1	2977.1	3076.5	3177.4
60.1°C	C_P		1.9222	1.9268	1.9469	1.9730	2.0021	2.0328
	S		8.1263	8.3680	8.5843	8.7811	8.9625	9.1312
	ρ		0.1750	0.1540	0.1376	0.1244	0.1135	0.1044
30.0 kPa	h		2685.0	2781.6	2878.7	2976.8	3076.2	3177.2
69.1°C	C_P		1.9390	1.9337	1.9504	1.9750	2.0033	2.0337
	S		7.9365	8.1796	8.3964	8.5935	8.7750	8.9438
	ρ		0.2337	0.2055	0.1836	0.1659	0.1514	0.1392
40.0 kPa	h		2683.7	2780.9	2878.2	2976.5	3076.0	3177.0
75.9°C	C_P		1.9564	1.9407	1.9539	1.9771	2.0046	2.0346
	S		7.8010	8.0456	8.2629	8.4603	8.6419	8.8108
	ρ		0.2925	0.2571	0.2296	0.2074	0.1893	0.1740
50.0 kPa	h		2682.4	2780.2	2877.8	2976.1	3075.8	3176.8
81.3°C	C_P		1.9743	1.9478	1.9574	1.9791	2.0059	2.0355
	S		7.6953	7.9413	8.1592	8.3568	8.5386	8.7076
	ρ		0.3516	0.3088	0.2756	0.2490	0.2272	0.2088
60.0 kPa	h		2681.1	2779.5	2877.3	2975.8	3075.5	3176.6
85.9°C	C_P		1.9928	1.9550	1.9609	1.9811	2.0072	2.0363
	S		7.6084	7.8558	8.0743	8.2722	8.4542	8.6232
	ρ		0.4702	0.4124	0.3679	0.3323	0.3030	0.2786
80.0 kPa	h		2678.5	2778.1	2876.4	2975.2	3075.0	3176.2
93.5°C	C_P		2.0322	1.9696	1.9681	1.9852	2.0098	2.0381
	S		7.4699	7.7204	7.9400	8.1385	8.3208	8.4900
	ρ		0.5897	0.5164	0.4603	0.4156	0.3790	0.3483
100.0 kPa	h		2675.8	2776.6	2875.5	2974.5	3074.5	3175.8
99.6°C	C_P		2.0766	1.9846	1.9754	1.9893	2.0124	2.0399
	S		7.3610	7.6148	7.8356	8.0346	8.2172	8.3866

T in °C; ρ in kg m^{-3}; h in kJ kg^{-1}; C_P in kJ kg^{-1} K^{-1}; S in kJ kg^{-1} K^{-1}

Superheated steam properties (cont.)

	T	400	500	600	700	800	900	1000
	ρ	0.00322	0.00280	0.00248	0.00223	0.00202	0.00185	0.00170
1.00 kPa	h	3280.1	3489.8	3706.3	3930.0	4160.7	4398.4	4642.8
7.0°C	C_P	2.0636	2.1309	2.2006	2.2716	2.3424	2.4114	2.4776
	S	10.672	10.963	11.226	11.468	11.694	11.906	12.106
	ρ	0.01610	0.01401	0.01241	0.01113	0.01010	0.00923	0.00851
5.00 kPa	h	3280.0	3489.7	3706.3	3929.9	4160.6	4398.4	4642.8
32.9°C	C_P	2.0639	2.1310	2.2007	2.2717	2.3424	2.4114	2.4776
	S	9.9293	10.220	10.483	10.725	10.951	11.163	11.363
	ρ	0.03219	0.02803	0.02482	0.02227	0.02019	0.01847	0.01702
10.0 kPa	h	3279.9	3489.7	3706.3	3929.9	4160.6	4398.3	4642.8
45.8°C	C_P	2.0642	2.1312	2.2008	2.2718	2.3425	2.4115	2.4777
	S	9.6094	9.8998	10.163	10.406	10.631	10.843	11.043
	ρ	0.06439	0.05606	0.04964	0.04453	0.04048	0.03694	0.03404
20.0 kPa	h	3279.8	3489.6	3706.2	3929.8	4160.6	4398.3	4642.8
60.1°C	C_P	2.0648	2.1316	2.2011	2.2719	2.3426	2.4115	2.4777
	S	9.2893	9.5798	9.8431	10.086	10.311	10.523	10.723
	ρ	0.09660	0.08409	0.07446	0.06680	0.06058	0.05541	0.05106
30.0 kPa	h	3279.6	3489.5	3706.1	3929.8	4160.5	4398.3	4642.8
69.1°C	C_P	2.0654	2.1319	2.2013	2.2721	2.3427	2.4116	2.4778
	S	9.1020	9.3925	9.6559	9.8984	10.124	10.336	10.536
	ρ	0.1288	0.1121	0.09928	0.08907	0.08077	0.07388	0.06808
40.0 kPa	h	3279.5	3489.4	3706.0	3929.7	4160.5	4398.2	4642.7
75.9°C	C_P	2.0661	2.1323	2.2015	2.2722	2.3428	2.4117	2.4779
	S	8.9691	9.2596	9.5231	9.7656	9.9912	10.203	10.403
	ρ	0.1611	0.1402	0.1241	0.1113	0.1010	0.09235	0.08510
50.0 kPa	h	3279.3	3489.3	3706.0	3929.7	4160.4	4398.2	4642.7
81.3°C	C_P	2.0667	2.1326	2.2017	2.2724	2.3429	2.4118	2.4779
	S	8.8659	9.1566	9.4201	9.6625	9.8882	10.100	10.300
	ρ	0.1933	0.1682	0.1489	0.1336	0.1212	0.1108	0.1021
60.0 kPa	h	3279.2	3489.2	3705.9	3929.6	4160.4	4398.2	4642.7
85.9°C	C_P	2.0673	2.1330	2.2020	2.2725	2.3430	2.4119	2.4780
	S	8.7816	9.0723	9.3358	9.5784	9.8040	10.016	10.216
	ρ	0.2578	0.2243	0.1986	0.1782	0.1616	0.1478	0.1362
80.0 kPa	h	3278.9	3488.9	3705.7	3929.5	4160.3	4398.1	4642.6
93.5°C	C_P	2.0686	2.1337	2.2024	2.2728	2.3432	2.4120	2.4781
	S	8.6485	8.9393	9.2029	9.4455	9.6712	9.8830	10.083
	ρ	0.3223	0.2805	0.2483	0.2227	0.2019	0.1847	0.1702
100.0 kPa	h	3278.6	3488.7	3705.6	3929.4	4160.2	4398.0	4642.6
99.6°C	C_P	2.0698	2.1344	2.2029	2.2731	2.3434	2.4122	2.4782
	S	8.5452	8.8361	9.0998	9.3424	9.5681	9.7800	9.9800

T in °C; ρ in kg m^{-3}; h in kJ kg^{-1}; C_P in kJ kg^{-1} K^{-1}; S in kJ kg^{-1} K^{-1}

7

Superheated steam properties (cont.)

	T	150	200	250	300	350	400	450
101.3 kPa	ρ	0.5233	0.4664	0.4211	0.3840	0.3529	0.3266	0.3039
	h	2776.5	2875.4	2974.5	3074.5	3175.8	3278.5	3382.8
100.0°C	C_P	1.9856	1.9759	1.9896	2.0126	2.0400	2.0699	2.1016
	S	7.6085	7.8294	8.0284	8.2110	8.3805	8.5391	8.6885
120 kPa	ρ	0.6207	0.5530	0.4991	0.4550	0.4181	0.3869	0.3600
	h	2775.1	2874.5	2973.9	3074.0	3175.4	3278.3	3382.6
104.8°C	C_P	2.0000	1.9828	1.9935	2.0150	2.0416	2.0711	2.1025
	S	7.5279	7.7499	7.9495	8.1324	8.3020	8.4607	8.6102
150 kPa	ρ	0.7779	0.6923	0.6245	0.5691	0.5229	0.4838	0.4501
	h	2772.9	2873.1	2972.9	3073.3	3174.9	3277.8	3382.2
111.3°C	C_P	2.0238	1.9940	1.9998	2.0190	2.0443	2.0730	2.1039
	S	7.4208	7.6447	7.8451	8.0284	8.1983	8.3572	8.5068
200 kPa	ρ	1.0418	0.9255	0.8341	0.7597	0.6979	0.6454	0.6004
	h	2769.1	2870.7	2971.2	3072.1	3173.9	3277.0	3381.6
120.2°C	C_P	2.0656	2.0133	2.0105	2.0256	2.0488	2.0762	2.1063
	S	7.2810	7.5081	7.7100	7.8941	8.0644	8.2236	8.3734
250 kPa	ρ	1.3082	1.1600	1.0445	0.9508	0.8731	0.8073	0.7509
	h	2765.2	2868.3	2969.5	3070.8	3172.9	3276.2	3380.9
127.4°C	C_P	2.1104	2.0332	2.0213	2.0323	2.0533	2.0794	2.1086
	S	7.1708	7.4012	7.6046	7.7895	7.9603	8.1197	8.2697
300 kPa	ρ	1.5773	1.3958	1.2556	1.1424	1.0486	0.9694	0.9015
	h	2761.2	2865.9	2967.9	3069.6	3172.0	3275.5	3380.3
133.5°C	C_P	2.1590	2.0537	2.0324	2.0391	2.0578	2.0826	2.1110
	S	7.0791	7.3131	7.5180	7.7037	7.8750	8.0347	8.1849
350 kPa	ρ	1.8490	1.6330	1.4675	1.3344	1.2245	1.1317	1.0523
	h	2757.1	2863.4	2966.2	3068.3	3171.0	3274.7	3379.7
138.9°C	C_P	2.2129	2.0750	2.0437	2.0460	2.0624	2.0859	2.1134
	S	7.0003	7.2380	7.4444	7.6309	7.8027	7.9627	8.1131
400 kPa	ρ	2.1237	1.8715	1.6801	1.5270	1.4006	1.2943	1.2032
	h	2752.8	2860.9	2964.5	3067.1	3170.0	3273.9	3379.0
143.6°C	C_P	2.2747	2.0969	2.0552	2.0529	2.0670	2.0891	2.1158
	S	6.9306	7.1723	7.3804	7.5677	7.7399	7.9002	8.0508
450 kPa	ρ	2.4014	2.1114	1.8936	1.7200	1.5771	1.4570	1.3543
	h	2748.3	2858.4	2962.8	3065.8	3169.1	3273.1	3378.4
147.9°C	C_P	2.3485	2.1195	2.0669	2.0599	2.0716	2.0924	2.1182
	S	6.8678	7.1138	7.3235	7.5117	7.6844	7.8450	7.9958
500 kPa	ρ		2.3528	2.1078	1.9135	1.7539	1.6199	1.5056
	h		2855.8	2961.0	3064.6	3168.1	3272.3	3377.7
151.8°C	C_P		2.1429	2.0788	2.0670	2.0763	2.0957	2.1206
	S		7.0610	7.2724	7.4614	7.6346	7.7955	7.9465

T in °C; ρ in kg m^{-3}; h in kJ kg^{-1}; C_P in kJ kg^{-1} K^{-1}; S in kJ kg^{-1} K^{-1}

Superheated steam properties (cont.)

	T	500	550	600	700	800	900	1000
	ρ	0.2842	0.2669	0.2516	0.2257	0.2046	0.1872	0.1725
101.3 kPa	h	3488.7	3596.3	3705.6	3929.4	4160.2	4398.0	4642.6
100.0°C	C_P	2.1345	2.1684	2.2029	2.2732	2.3435	2.4122	2.4782
	S	8.8301	8.9649	9.0937	9.3363	9.5621	9.7739	9.9739
	ρ	0.3366	0.3161	0.2980	0.2673	0.2423	0.2217	0.2042
120.0 kPa	h	3488.5	3596.1	3705.4	3929.3	4160.1	4397.9	4642.5
104.8°C	C_P	2.1352	2.1689	2.2033	2.2735	2.3437	2.4124	2.4784
	S	8.7518	8.8866	9.0155	9.2582	9.4839	9.6958	9.8958
	ρ	0.4209	0.3952	0.3725	0.3341	0.3029	0.2771	0.2553
150.0 kPa	h	3488.2	3595.8	3705.2	3929.1	4160.0	4397.8	4642.4
111.3°C	C_P	2.1363	2.1697	2.2040	2.2739	2.3440	2.4126	2.4786
	S	8.6485	8.7834	8.9124	9.1550	9.3808	9.5927	9.7927
	ρ	0.5614	0.5271	0.4968	0.4456	0.4040	0.3695	0.3404
200.0 kPa	h	3487.7	3595.4	3704.8	3928.8	4159.8	4397.6	4642.3
120.2°C	C_P	2.1381	2.1712	2.2052	2.2747	2.3445	2.4130	2.4789
	S	8.5152	8.6502	8.7792	9.0220	9.2479	9.4598	9.6599
	ρ	0.7020	0.6590	0.6211	0.5570	0.5050	0.4619	0.4255
250.0 kPa	h	3487.1	3594.9	3704.4	3928.5	4159.5	4397.5	4642.1
127.4°C	C_P	2.1399	2.1726	2.2063	2.2755	2.3451	2.4134	2.4792
	S	8.4117	8.5468	8.6759	8.9188	9.1448	9.3567	9.5568
	ρ	0.8427	0.7911	0.7455	0.6685	0.6060	0.5543	0.5107
300.0 kPa	h	3486.6	3594.5	3704.0	3928.2	4159.3	4397.3	4642.0
133.5°C	C_P	2.1417	2.1740	2.2075	2.2762	2.3456	2.4138	2.4795
	S	8.3271	8.4623	8.5914	8.8344	9.0604	9.2724	9.4726
	ρ	0.9835	0.9232	0.8699	0.7801	0.7071	0.6467	0.5958
350.0 kPa	h	3486.1	3594.0	3703.6	3927.9	4159.1	4397.1	4641.8
138.9°C	C_P	2.1435	2.1755	2.2086	2.2770	2.3462	2.4142	2.4798
	S	8.2554	8.3907	8.5200	8.7631	8.9891	9.2012	9.4013
	ρ	1.1244	1.0554	0.9945	0.8916	0.8082	0.7391	0.6809
400.0 kPa	h	3485.5	3593.6	3703.2	3927.6	4158.8	4396.9	4641.7
143.6°C	C_P	2.1454	2.1769	2.2098	2.2778	2.3467	2.4147	2.4801
	S	8.1933	8.3287	8.4580	8.7012	8.9273	9.1394	9.3396
	ρ	1.2654	1.1877	1.1190	1.0032	0.9093	0.8315	0.7661
450.0 kPa	h	3485.0	3593.1	3702.9	3927.3	4158.6	4396.7	4641.5
147.9°C	C_P	2.1472	2.1783	2.2109	2.2786	2.3473	2.4151	2.4804
	S	8.1384	8.2739	8.4033	8.6466	8.8728	9.0849	9.2851
	ρ	1.4066	1.3200	1.2436	1.1149	1.0104	0.9240	0.8512
500.0 kPa	h	3484.5	3592.7	3702.5	3927.0	4158.4	4396.6	4641.4
151.8°C	C_P	2.1490	2.1798	2.2121	2.2793	2.3479	2.4155	2.4807
	S	8.0892	8.2249	8.3543	8.5977	8.8240	9.0362	9.2364

T in °C; ρ in kg m^{-3}; h in kJ kg^{-1}; C_P in kJ kg^{-1} K^{-1}; S in kJ kg^{-1} K^{-1}

Superheated steam properties (cont.)

	T	200	250	300	350	400	450	500
	ρ	2.8399	2.5387	2.3019	2.1085	1.9465	1.8085	1.6892
600 kPa	h	2850.6	2957.6	3062.0	3166.1	3270.8	3376.5	3483.4
158.8°C	C_P	2.1920	2.1033	2.0815	2.0857	2.1023	2.1255	2.1527
	S	6.9683	7.1832	7.3740	7.5481	7.7097	7.8611	8.0041
	ρ	3.3333	2.9729	2.6924	2.4643	2.2739	2.1120	1.9722
700 kPa	h	2845.3	2954.0	3059.4	3164.2	3269.2	3375.2	3482.3
164.9°C	C_P	2.2446	2.1287	2.0962	2.0953	2.1089	2.1303	2.1564
	S	6.8884	7.1070	7.2995	7.4746	7.6368	7.7886	7.9319
	ρ	3.8332	3.4106	3.0849	2.8215	2.6022	2.4161	2.2557
800 kPa	h	2839.7	2950.4	3056.9	3162.2	3267.6	3373.9	3481.3
170.4°C	C_P	2.3009	2.1550	2.1113	2.1050	2.1157	2.1353	2.1602
	S	6.8176	7.0401	7.2345	7.4106	7.5734	7.7257	7.8692
	ρ	4.3399	3.8517	3.4795	3.1800	2.9314	2.7208	2.5396
900 kPa	h	2834.1	2946.8	3054.3	3160.2	3266.1	3372.6	3480.2
175.4°C	C_P	2.3618	2.1823	2.1267	2.1148	2.1224	2.1402	2.1639
	S	6.7539	6.9805	7.1767	7.3539	7.5173	7.6700	7.8138
	ρ	4.8539	4.2965	3.8762	3.5398	3.2615	3.0262	2.8240
1000 kPa	h	2828.3	2943.1	3051.6	3158.2	3264.5	3371.3	3479.1
179.9°C	C_P	2.4281	2.2106	2.1425	2.1248	2.1293	2.1452	2.1677
	S	6.6955	6.9265	7.1246	7.3029	7.4669	7.6200	7.7641
	ρ	6.1735	5.4250	4.8773	4.4453	4.0907	3.7923	3.5368
1250 kPa	h	2812.9	2933.7	3045.0	3153.1	3260.5	3368.1	3476.4
189.8°C	C_P	2.6272	2.2860	2.1834	2.1503	2.1467	2.1577	2.1771
	S	6.5667	6.8096	7.0129	7.1939	7.3596	7.5137	7.6585
	ρ	7.5498	6.5785	5.8925	5.3594	4.9256	4.5624	4.2524
1500 kPa	h	2796.0	2923.9	3038.2	3148.0	3256.5	3364.8	3473.7
198.3°C	C_P	2.9091	2.3685	2.2266	2.1768	2.1645	2.1705	2.1867
	S	6.4536	6.7111	6.9198	7.1036	7.2710	7.4262	7.5718
	ρ		7.7590	6.9224	6.2826	5.7664	5.3365	4.9708
1750 kPa	h		2913.8	3031.3	3142.9	3252.4	3361.5	3471.0
205.7°C	C_P		2.4589	2.2722	2.2041	2.1827	2.1835	2.1964
	S		6.6248	6.8395	7.0262	7.1953	7.3516	7.4980
	ρ		8.9689	7.9677	7.2150	6.6131	6.1146	5.6921
2000 kPa	h		2903.2	3024.2	3137.7	3248.3	3358.2	3468.2
212.4°C	C_P		2.5584	2.3203	2.2324	2.2013	2.1966	2.2062
	S		6.5475	6.7684	6.9583	7.1292	7.2866	7.4337
	ρ		11.487	10.108	9.1087	8.3251	7.6833	7.1433
2500 kPa	h		2880.9	3009.6	3127.0	3240.1	3351.6	3462.7
223.9°C	C_P		2.7903	2.4248	2.2920	2.2398	2.2236	2.2260
	S		6.4107	6.6459	6.8424	7.0170	7.1767	7.3254

T in °C; ρ in kg m^{-3}; h in kJ kg^{-1}; C_P in kJ kg^{-1} K^{-1}; S in kJ kg^{-1} K^{-1}

Superheated steam properties (cont.)

	T	550	600	650	700	800	900	1000
	ρ	1.5850	1.4931	1.4114	1.3382	1.2128	1.1089	1.0215
600 kPa	h	3591.8	3701.7	3813.2	3926.4	4157.9	4396.2	4641.1
158.8°C	C_P	2.1827	2.2144	2.2473	2.2809	2.3490	2.4163	2.4814
	S	8.1399	8.2695	8.3937	8.5131	8.7395	8.9518	9.1521
	ρ	1.8502	1.7427	1.6472	1.5617	1.4151	1.2939	1.1919
700 kPa	h	3590.9	3700.9	3812.6	3925.8	4157.5	4395.8	4640.8
164.9°C	C_P	2.1856	2.2167	2.2492	2.2825	2.3501	2.4171	2.4820
	S	8.0679	8.1977	8.3220	8.4415	8.6680	8.8804	9.0807
	ρ	2.1158	1.9926	1.8832	1.7854	1.6176	1.4789	1.3622
800 kPa	h	3590.0	3700.1	3811.9	3925.3	4157.0	4395.5	4640.5
170.4°C	C_P	2.1885	2.2190	2.2510	2.2840	2.3512	2.4179	2.4826
	S	8.0054	8.1354	8.2598	8.3794	8.6061	8.8185	9.0189
	ρ	2.3817	2.2427	2.1194	2.0091	1.8201	1.6639	1.5326
900 kPa	h	3589.0	3699.4	3811.2	3924.7	4156.6	4395.1	4640.2
175.4°C	C_P	2.1914	2.2213	2.2529	2.2856	2.3523	2.4188	2.4832
	S	7.9503	8.0803	8.2049	8.3246	8.5514	8.7639	8.9643
	ρ	2.6479	2.4931	2.3557	2.2330	2.0227	1.8490	1.7030
1000 kPa	h	3588.1	3698.6	3810.5	3924.1	4156.1	4394.8	4639.9
179.9°C	C_P	2.1943	2.2237	2.2548	2.2871	2.3534	2.4196	2.4839
	S	7.9008	8.0310	8.1557	8.2755	8.5024	8.7150	8.9155
	ρ	3.3148	3.1200	2.9474	2.7933	2.5296	2.3119	2.1290
1250 kPa	h	3585.9	3696.6	3808.8	3922.6	4155.0	4393.8	4639.2
189.8°C	C_P	2.2016	2.2295	2.2596	2.2911	2.3562	2.4216	2.4854
	S	7.7957	7.9263	8.0513	8.1713	8.3985	8.6113	8.8120
	ρ	3.9838	3.7484	3.5401	3.3544	3.0368	2.7750	2.5553
1500 kPa	h	3583.6	3694.7	3807.1	3921.1	4153.8	4392.9	4638.5
198.3°C	C_P	2.2091	2.2354	2.2643	2.2950	2.3590	2.4237	2.4870
	S	7.7095	7.8405	7.9658	8.0860	8.3135	8.5266	8.7274
	ρ	4.6548	4.3783	4.1340	3.9163	3.5446	3.2384	2.9816
1750 kPa	h	3581.3	3692.7	3805.5	3919.6	4152.7	4392.0	4637.8
205.7°C	C_P	2.2165	2.2413	2.2691	2.2989	2.3617	2.4257	2.4886
	S	7.6362	7.7676	7.8932	8.0137	8.2415	8.4548	8.6557
	ρ	5.3278	5.0097	4.7289	4.4790	4.0528	3.7021	3.4081
2000 kPa	h	3579.0	3690.7	3803.8	3918.2	4151.5	4391.1	4637.0
212.4°C	C_P	2.2241	2.2473	2.2740	2.3029	2.3645	2.4278	2.4902
	S	7.5725	7.7043	7.8302	7.9509	8.1790	8.3925	8.5936
	ρ	6.6800	6.2769	5.9221	5.6070	5.0706	4.6302	4.2615
2500 kPa	h	3574.3	3686.8	3800.4	3915.2	4149.2	4389.3	4635.6
223.9°C	C_P	2.2393	2.2593	2.2837	2.3109	2.3701	2.4319	2.4933
	S	7.4653	7.5979	7.7243	7.8455	8.0743	8.2882	8.4896

T in °C; ρ in kg m^{-3}; h in kJ kg^{-1}; C_P in kJ kg^{-1} K^{-1}; S in kJ kg^{-1} K^{-1}

11

Superheated steam properties (cont.)

	T	250	300	350	400	450	500	550
	ρ	14.159	12.318	11.043	10.062	9.2690	8.6060	8.0410
3000 kPa	h	2856.5	2994.3	3116.1	3231.7	3344.8	3457.2	3569.7
233.9°C	C_P	3.0831	2.5414	2.3559	2.2801	2.2514	2.2464	2.2548
	S	6.2893	6.5412	6.7449	6.9234	7.0856	7.2359	7.3768
	ρ	17.019	14.609	13.020	11.826	10.872	10.081	9.4100
3500 kPa	h	2829.7	2978.4	3104.8	3223.2	3338.0	3451.6	3565.0
242.6°C	C_P	3.4738	2.6719	2.4243	2.3223	2.2801	2.2672	2.2706
	S	6.1764	6.4484	6.6601	6.8427	7.0074	7.1593	7.3014
	ρ		16.987	15.044	13.618	12.493	11.568	10.788
4000 kPa	h		2961.7	3093.3	3214.5	3331.2	3446.0	3560.3
250.4°C	C_P		2.8185	2.4976	2.3665	2.3097	2.2884	2.2865
	S		6.3639	6.5843	6.7714	6.9386	7.0922	7.2355
	ρ		19.464	17.117	15.439	14.133	13.068	12.174
4500 kPa	h		2944.2	3081.5	3205.6	3324.2	3440.4	3555.6
257.4°C	C_P		2.9840	2.5763	2.4127	2.3402	2.3101	2.3028
	S		6.2854	6.5153	6.7070	6.8770	7.0323	7.1767
	ρ		22.053	19.242	17.290	15.792	14.581	13.570
5000 kPa	h		2925.7	3069.3	3196.7	3317.2	3434.7	3550.9
263.9°C	C_P		3.1722	2.6608	2.4610	2.3717	2.3323	2.3193
	S		6.2110	6.4516	6.6483	6.8210	6.9781	7.1237
	ρ		24.769	21.425	19.173	17.471	16.107	14.974
5500 kPa	h		2906.2	3056.8	3187.5	3310.1	3428.9	3546.1
270.0°C	C_P		3.3881	2.7517	2.5117	2.4041	2.3550	2.3361
	S		6.1397	6.3920	6.5939	6.7696	6.9285	7.0753
	ρ		27.632	23.668	21.088	19.170	17.646	16.388
6000 kPa	h		2885.5	3043.9	3178.2	3302.9	3423.1	3541.3
275.6°C	C_P		3.6388	2.8497	2.5647	2.4376	2.3782	2.3531
	S		6.0703	6.3357	6.5432	6.7219	6.8826	7.0307
	ρ		30.668	25.978	23.039	20.890	19.199	17.811
6500 kPa	h		2863.5	3030.6	3168.8	3295.6	3417.3	3536.5
280.9°C	C_P		3.9343	2.9554	2.6204	2.4722	2.4019	2.3704
	S		6.0019	6.2820	6.4954	6.6773	6.8399	6.9893
	ρ		33.907	28.359	25.026	22.631	20.765	19.244
7000 kPa	h		2839.9	3016.9	3159.2	3288.3	3411.4	3531.6
285.8°C	C_P		4.2898	3.0698	2.6786	2.5079	2.4261	2.3880
	S		5.9337	6.2304	6.4502	6.6353	6.8000	6.9506
	ρ		37.394	30.818	27.052	24.395	22.346	20.686
7500 kPa	h		2814.4	3002.8	3149.4	3280.9	3405.5	3526.7
290.5°C	C_P		4.7294	3.1938	2.7398	2.5447	2.4509	2.4059
	S		5.8646	6.1806	6.4071	6.5956	6.7623	6.9143

T in °C; $\qquad$ ρ in kg m^{-3}; $\qquad$ h in kJ kg^{-1}; $\qquad$ C_P in kJ kg^{-1} K^{-1}; $\qquad$ S in kJ kg^{-1} K^{-1}

Superheated steam properties (cont.)

	T	600	650	700	750	800	900	1000
	ρ	7.5500	7.1200	6.7380	6.3970	6.0900	5.5590	5.1150
3000 kPa	h	3682.8	3796.9	3912.2	4028.9	4146.9	4387.5	4634.1
233.9°C	C_P	2.2715	2.2934	2.3189	2.3466	2.3758	2.4361	2.4965
	S	7.5103	7.6373	7.7590	7.8758	7.9885	8.2028	8.4045
	ρ	8.8300	8.3220	7.8730	7.4720	7.1120	6.4890	5.9700
3500 kPa	h	3678.9	3793.5	3909.3	4026.3	4144.6	4385.7	4632.7
242.6°C	C_P	2.2838	2.3033	2.3270	2.3533	2.3814	2.4402	2.4996
	S	7.4356	7.5633	7.6854	7.8027	7.9156	8.1303	8.3324
	ρ	10.115	9.5290	9.0110	8.5500	8.1350	7.4210	6.8250
4000 kPa	h	3674.9	3790.1	3906.3	4023.6	4142.3	4383.9	4631.2
250.4°C	C_P	2.2963	2.3133	2.3351	2.3601	2.3871	2.4444	2.5028
	S	7.3705	7.4988	7.6214	7.7390	7.8523	8.0674	8.2697
	ρ	11.407	10.740	10.152	9.6294	9.1605	8.3528	7.6803
4500 kPa	h	3670.9	3786.6	3903.3	4021.0	4140.0	4382.1	4629.8
257.4°C	C_P	2.3089	2.3233	2.3433	2.3669	2.3928	2.4486	2.5059
	S	7.3127	7.4416	7.5646	7.6826	7.7962	8.0118	8.2144
	ρ	12.706	11.956	11.297	10.712	10.188	9.2860	8.5360
5000 kPa	h	3666.8	3783.2	3900.3	4018.4	4137.7	4380.2	4628.3
263.9°C	C_P	2.3216	2.3335	2.3515	2.3737	2.3986	2.4528	2.5091
	S	7.2605	7.3901	7.5136	7.6320	7.7458	7.9618	8.1648
	ρ	14.011	13.177	12.445	11.797	11.217	10.220	9.3930
5500 kPa	h	3662.8	3779.7	3897.3	4015.8	4135.4	4378.4	4626.9
270.0°C	C_P	2.3345	2.3437	2.3598	2.3806	2.4043	2.4569	2.5123
	S	7.2130	7.3432	7.4672	7.5859	7.7001	7.9166	8.1198
	ρ	15.322	14.402	13.597	12.884	12.248	11.156	10.250
6000 kPa	h	3658.7	3776.2	3894.3	4013.2	4133.1	4376.6	4625.4
275.6°C	C_P	2.3476	2.3540	2.3682	2.3874	2.4101	2.4612	2.5155
	S	7.1693	7.3001	7.4246	7.5438	7.6582	7.8751	8.0786
	ρ	16.640	15.632	14.752	13.974	13.281	12.092	11.108
6500 kPa	h	3654.7	3772.8	3891.3	4010.5	4130.8	4374.8	4624.0
280.9°C	C_P	2.3608	2.3644	2.3766	2.3944	2.4159	2.4654	2.5187
	S	7.1288	7.2603	7.3853	7.5048	7.6195	7.8369	8.0407
	ρ	17.965	16.867	15.911	15.067	14.315	13.029	11.966
7000 kPa	h	3650.6	3769.3	3888.2	4007.9	4128.4	4373.0	4622.5
285.8°C	C_P	2.3742	2.3749	2.3851	2.4013	2.4217	2.4696	2.5219
	S	7.0910	7.2231	7.3486	7.4685	7.5836	7.8014	8.0055
	ρ	19.296	18.107	17.073	16.162	15.352	13.968	12.825
7500 kPa	h	3646.5	3765.8	3885.2	4005.2	4126.1	4371.1	4621.1
290.5°C	C_P	2.3877	2.3855	2.3936	2.4083	2.4275	2.4738	2.5251
	S	7.0555	7.1884	7.3144	7.4346	7.5500	7.7682	7.9726

T in °C; ρ in kg m^{-3}; h in kJ kg^{-1}; C_P in kJ kg^{-1} K^{-1}; S in kJ kg^{-1} K^{-1}

13

Superheated steam properties (cont.)

	T	350	400	450	500	550	600	650
	ρ	33.362	29.117	26.182	23.942	22.138	20.634	19.352
8000 kPa	h	2988.1	3139.4	3273.3	3399.5	3521.8	3642.4	3762.3
295.0°C	C_P	3.3287	2.8040	2.5827	2.4762	2.4240	2.4014	2.3962
	S	6.1321	6.3658	6.5579	6.7266	6.8799	7.0221	7.1556
	ρ	35.998	31.225	27.993	25.553	23.600	21.979	20.601
8500 kPa	h	2973.0	3129.2	3265.7	3393.5	3516.9	3638.2	3758.7
299.3°C	C_P	3.4760	2.8714	2.6219	2.5020	2.4424	2.4152	2.4070
	S	6.0846	6.3260	6.5218	6.6927	6.8474	6.9905	7.1247
	ρ	38.736	33.378	29.829	27.179	25.072	23.331	21.856
9000 kPa	h	2957.3	3118.8	3258.0	3387.4	3512.0	3634.1	3755.2
303.3°C	C_P	3.6374	2.9423	2.6623	2.5284	2.4611	2.4292	2.4179
	S	6.0380	6.2876	6.4872	6.6603	6.8164	6.9605	7.0953
	ρ	44.564	37.827	33.578	30.478	28.047	26.057	24.380
10,000 kPa	h	2924.0	3097.4	3242.3	3375.1	3502.0	3625.8	3748.1
311.0°C	C_P	4.0117	3.0953	2.7473	2.5830	2.4994	2.4576	2.4399
	S	5.9459	6.2141	6.4219	6.5995	6.7585	6.9045	7.0408
	ρ	50.955	42.486	37.435	33.842	31.064	28.811	26.924
11,000 kPa	h	2887.9	3075.2	3226.3	3362.7	3491.9	3617.4	3741.0
318.1°C	C_P	4.4760	3.2652	2.8379	2.6401	2.5389	2.4867	2.4623
	S	5.8542	6.1440	6.3607	6.5431	6.7050	6.8531	6.9908
	ρ	58.068	47.380	41.410	37.275	34.125	31.594	29.488
12,000 kPa	h	2848.1	3052.0	3209.8	3350.0	3481.7	3608.9	3733.8
324.7°C	C_P	5.0703	3.4548	2.9346	2.6996	2.5797	2.5165	2.4850
	S	5.7609	6.0764	6.3028	6.4903	6.6553	6.8054	6.9446
	ρ	66.144	52.541	45.509	40.779	37.231	34.408	32.074
13,000 kPa	h	2803.7	3027.7	3193.0	3337.1	3471.3	3600.4	3726.6
330.9°C	C_P	5.8653	3.6674	3.0380	2.7618	2.6217	2.5469	2.5082
	S	5.6638	6.0106	6.2476	6.4405	6.6087	6.7609	6.9016
	ρ	75.577	58.003	49.743	44.357	40.383	37.252	34.680
14,000 kPa	h	2753.1	3002.3	3175.7	3324.1	3460.9	3591.8	3719.4
336.7°C	C_P	7.0019	3.9071	3.1486	2.8268	2.6649	2.5780	2.5317
	S	5.5598	5.9459	6.1946	6.3932	6.5648	6.7191	6.8612
	ρ	87.100	63.812	54.121	48.014	43.583	40.127	37.308
15,000 kPa	h	2693.1	2975.7	3157.9	3310.8	3450.4	3583.1	3712.1
342.2°C	C_P	8.8167	4.1793	3.2670	2.8947	2.7095	2.6098	2.5555
	S	5.4437	5.8819	6.1434	6.3480	6.5230	6.6796	6.8233
	ρ	102.40	70.021	58.654	51.752	46.832	43.034	39.957
16,000 kPa	h	2617.0	2947.6	3139.7	3297.3	3439.8	3574.4	3704.8
347.4°C	C_P	12.424	4.4908	3.3939	2.9657	2.7554	2.6422	2.5798
	S	5.3045	5.8179	6.0937	6.3046	6.4832	6.6421	6.7873

T in °C; ρ in kg m^{-3}; h in kJ kg^{-1}; C_P in kJ kg^{-1} K^{-1}; S in kJ kg^{-1} K^{-1}

Superheated steam properties (cont.)

	T	700	750	800	850	900	950	1000
	ρ	18.239	17.260	16.391	15.611	14.907	14.268	13.684
8000 kPa	h	3882.2	4002.6	4123.8	4246.0	4369.3	4493.8	4619.6
295.0°C	C_P	2.4022	2.4154	2.4334	2.4547	2.4781	2.5028	2.5283
	S	7.2821	7.4028	7.5184	7.6297	7.7371	7.8411	7.9419
	ρ	19.408	18.361	17.431	16.599	15.848	15.166	14.544
8500 kPa	h	3879.1	3999.9	4121.5	4243.9	4367.5	4492.2	4618.1
299.3°C	C_P	2.4108	2.4224	2.4393	2.4597	2.4823	2.5065	2.5315
	S	7.2517	7.3728	7.4887	7.6002	7.7079	7.8120	7.9129
	ρ	20.581	19.464	18.473	17.588	16.789	16.065	15.404
9000 kPa	h	3876.1	3997.3	4119.1	4241.9	4365.7	4490.6	4616.7
303.3°C	C_P	2.4195	2.4295	2.4452	2.4646	2.4866	2.5102	2.5347
	S	7.2229	7.3443	7.4606	7.5724	7.6802	7.7844	7.8855
	ρ	22.937	21.678	20.564	19.570	18.675	17.865	17.126
10,000 kPa	h	3870.0	3992.0	4114.5	4237.8	4362.0	4487.3	4613.8
311.0°C	C_P	2.4370	2.4438	2.4571	2.4747	2.4952	2.5175	2.5411
	S	7.1693	7.2916	7.4085	7.5207	7.6290	7.7335	7.8349
	ρ	25.308	23.902	22.662	21.558	20.565	19.668	18.850
11,000 kPa	h	3863.9	3986.7	4109.8	4233.6	4358.3	4484.1	4610.9
318.1°C	C_P	2.4548	2.4583	2.4691	2.4847	2.5037	2.5250	2.5476
	S	7.1204	7.2434	7.3610	7.4737	7.5824	7.6873	7.7889
	ρ	27.694	26.137	24.768	23.551	22.460	21.473	20.577
12,000 kPa	h	3857.7	3981.3	4105.1	4229.5	4354.7	4480.8	4608.0
324.7°C	C_P	2.4728	2.4729	2.4811	2.4949	2.5124	2.5324	2.5540
	S	7.0753	7.1991	7.3173	7.4306	7.5396	7.6449	7.7467
	ρ	30.094	28.383	26.882	25.550	24.358	23.282	22.305
13,000 kPa	h	3851.5	3976.0	4100.4	4225.4	4351.0	4477.5	4605.0
330.9°C	C_P	2.4910	2.4876	2.4933	2.5051	2.5210	2.5398	2.5605
	S	7.0333	7.1580	7.2768	7.3906	7.5000	7.6056	7.7078
	ρ	32.509	30.639	29.003	27.555	26.261	25.094	24.035
14,000 kPa	h	3845.3	3970.6	4095.8	4221.3	4347.4	4474.3	4602.1
336.7°C	C_P	2.5094	2.5025	2.5055	2.5153	2.5298	2.5473	2.5670
	S	6.9941	7.1196	7.2391	7.3534	7.4632	7.5692	7.6716
	ρ	34.939	32.906	31.132	29.566	28.167	26.908	25.768
15,000 kPa	h	3839.1	3965.2	4091.1	4217.1	4343.7	4471.0	4599.2
342.2°C	C_P	2.5281	2.5174	2.5178	2.5256	2.5385	2.5548	2.5735
	S	6.9572	7.0836	7.2037	7.3185	7.4288	7.5350	7.6378
	ρ	37.385	35.184	33.269	31.582	30.078	28.726	27.502
16,000 kPa	h	3832.9	3959.8	4086.3	4213.0	4340.0	4467.8	4596.3
347.4°C	C_P	2.5470	2.5326	2.5302	2.5360	2.5473	2.5623	2.5800
	S	6.9224	7.0496	7.1703	7.2857	7.3964	7.5030	7.6060

T in °C; ρ in kg m^{-3}; h in kJ kg^{-1}; C_P in kJ kg^{-1} K^{-1}; S in kJ kg^{-1} K^{-1}

15

Superheated steam properties (cont.)

	T	400	425	450	500	550	600	650
	ρ	76.697	68.910	63.356	55.575	50.131	45.973	42.628
17,000 kPa	h	2917.9	3027.3	3121.0	3283.6	3429.0	3565.7	3697.5
352.3 °C	C_P	4.8508	4.0007	3.5302	3.0398	2.8028	2.6753	2.6044
	S	5.7536	5.9133	6.0451	6.2628	6.4451	6.6063	6.7531
	ρ	83.924	74.643	68.239	59.488	53.482	48.945	45.321
18,000 kPa	h	2886.4	3003.6	3101.8	3269.7	3418.2	3556.8	3690.1
357.0 °C	C_P	5.2713	4.2286	3.6766	3.1172	2.8515	2.7091	2.6294
	S	5.6883	5.8593	5.9975	6.2223	6.4084	6.5720	6.7205
	ρ	87.777	77.630	70.753	61.479	55.178	50.444	46.676
18,500 kPa	h	2869.9	2991.3	3091.9	3262.6	3412.7	3552.4	3686.4
359.3 °C	C_P	5.5093	4.3517	3.7540	3.1572	2.8764	2.7263	2.6420
	S	5.6551	5.8323	5.9740	6.2025	6.3906	6.5554	6.7047
	ρ	91.810	80.702	73.318	63.494	56.887	51.951	48.037
19,000 kPa	h	2852.8	2978.8	3082.0	3255.5	3407.2	3548.0	3682.7
361.5 °C	C_P	5.7691	4.4815	3.8342	3.1980	2.9016	2.7436	2.6548
	S	5.6215	5.8054	5.9506	6.1829	6.3731	6.5391	6.6892
	ρ	96.044	83.867	75.936	65.533	58.610	53.467	49.403
19,500 kPa	h	2835.2	2966.1	3071.9	3248.4	3401.7	3543.5	3679.0
363.6 °C	C_P	6.0539	4.6185	3.9176	3.2398	2.9272	2.7612	2.6676
	S	5.5873	5.7784	5.9274	6.1637	6.3559	6.5232	6.6741
	ρ	100.50	87.129	78.609	67.598	60.346	54.991	50.775
20,000 kPa	h	2816.9	2953.0	3061.7	3241.2	3396.1	3539.0	3675.3
365.7 °C	C_P	6.3675	4.7633	4.0041	3.2825	2.9532	2.7788	2.6805
	S	5.5525	5.7514	5.9043	6.1446	6.3389	6.5075	6.6593
	ρ	105.20	90.494	81.341	69.688	62.097	56.524	52.153
20,500 kPa	h	2797.8	2939.7	3051.3	3233.9	3390.6	3534.5	3671.5
367.8 °C	C_P	6.7146	4.9165	4.0939	3.3261	2.9796	2.7967	2.6935
	S	5.5171	5.7242	5.8813	6.1258	6.3222	6.4921	6.6447
	ρ	110.18	93.970	84.132	71.804	63.862	58.065	53.536
21,000 kPa	h	2778.0	2926.1	3040.7	3226.6	3385.0	3530.0	3667.8
369.8 °C	C_P	7.1009	5.0788	4.1873	3.3707	3.0063	2.8147	2.7066
	S	5.4808	5.6970	5.8584	6.1072	6.3058	6.4769	6.6304
	ρ	115.48	97.564	86.987	73.946	65.641	59.616	54.925
21,500 kPa	h	2757.4	2912.2	3030.1	3219.2	3379.3	3525.5	3664.1
371.8 °C	C_P	7.5334	5.2509	4.2844	3.4162	3.0334	2.8329	2.7198
	S	5.4435	5.6696	5.8355	6.0888	6.2896	6.4620	6.6163
	ρ	121.13	101.28	89.907	76.116	67.435	61.175	56.319
22,000 kPa	h	2735.8	2898.0	3019.2	3211.8	3373.7	3521.0	3660.3
373.7 °C	C_P	8.0211	5.4338	4.3854	3.4628	3.0609	2.8513	2.7330
	S	5.4051	5.6420	5.8127	6.0705	6.2736	6.4473	6.6025

T in °C; ρ in kg m^{-3}; h in kJ kg^{-1}; C_P in kJ kg^{-1} K^{-1}; S in kJ kg^{-1} K^{-1}

Superheated steam properties (cont.)

	T	700	750	800	850	900	950	1000
	ρ	39.845	37.473	35.414	33.603	31.992	30.546	29.238
17,000 kPa	h	3826.6	3954.4	4081.6	4208.8	4336.4	4464.5	4593.4
352.3°C	C_P	2.5660	2.5478	2.5427	2.5464	2.5561	2.5699	2.5866
	S	6.8894	7.0174	7.1388	7.2547	7.3658	7.4727	7.5760
	ρ	42.321	39.772	37.566	35.630	33.911	32.369	30.977
18,000 kPa	h	3820.4	3949.0	4076.9	4204.7	4332.7	4461.2	4590.5
357.0°C	C_P	2.5853	2.5632	2.5552	2.5568	2.5649	2.5774	2.5931
	S	6.8579	6.9868	7.1089	7.2252	7.3368	7.4440	7.5476
	ρ	43.564	40.926	38.646	36.646	34.871	33.282	31.847
18,500 kPa	h	3817.2	3946.3	4074.6	4202.6	4330.9	4459.6	4589.0
359.3°C	C_P	2.5950	2.5709	2.5615	2.5620	2.5693	2.5812	2.5964
	S	6.8427	6.9720	7.0944	7.2111	7.3228	7.4303	7.5340
	ρ	44.812	42.082	39.727	37.663	35.833	34.195	32.717
19,000 kPa	h	3814.1	3943.6	4072.2	4200.5	4329.0	4458.0	4587.6
361.5°C	C_P	2.6048	2.5787	2.5679	2.5673	2.5737	2.5850	2.5996
	S	6.8278	6.9576	7.0803	7.1972	7.3092	7.4168	7.5207
	ρ	46.063	43.241	40.810	38.681	36.796	35.109	33.588
19,500 kPa	h	3810.9	3940.9	4069.8	4198.5	4327.2	4456.4	4586.1
363.6°C	C_P	2.6147	2.5865	2.5742	2.5726	2.5782	2.5888	2.6029
	S	6.8133	6.9435	7.0666	7.1837	7.2959	7.4037	7.5077
	ρ	47.318	44.403	41.895	39.701	37.759	36.024	34.459
20,000 kPa	h	3807.8	3938.1	4067.5	4196.4	4325.4	4454.7	4584.7
365.7°C	C_P	2.6245	2.5943	2.5806	2.5778	2.5826	2.5926	2.6062
	S	6.7990	6.9297	7.0531	7.1705	7.2829	7.3909	7.4950
	ρ	48.577	45.568	42.981	40.722	38.724	36.939	35.331
20,500 kPa	h	3804.6	3935.4	4065.1	4194.3	4323.5	4453.1	4583.2
367.8°C	C_P	2.6345	2.6021	2.5869	2.5831	2.5871	2.5964	2.6095
	S	6.7851	6.9162	7.0399	7.1576	7.2702	7.3783	7.4826
	ρ	49.840	46.735	44.070	41.745	39.690	37.855	36.203
21,000 kPa	h	3801.4	3932.7	4062.7	4192.2	4321.7	4451.5	4581.8
369.8°C	C_P	2.6444	2.6100	2.5933	2.5884	2.5915	2.6002	2.6128
	S	6.7714	6.9029	7.0270	7.1450	7.2577	7.3661	7.4705
	ρ	51.107	47.905	45.161	42.768	40.656	38.772	37.075
21,500 kPa	h	3798.3	3930.0	4060.4	4190.2	4319.9	4449.9	4580.3
371.8°C	C_P	2.6545	2.6179	2.5997	2.5937	2.5960	2.6040	2.6161
	S	6.7579	6.8899	7.0144	7.1326	7.2456	7.3541	7.4586
	ρ	52.378	49.078	46.253	43.794	41.624	39.689	37.948
22,000 kPa	h	3795.1	3927.3	4058.0	4188.1	4318.0	4448.2	4578.9
373.7°C	C_P	2.6646	2.6259	2.6062	2.5990	2.6005	2.6079	2.6194
	S	6.7447	6.8772	7.0020	7.1204	7.2336	7.3423	7.4470

T in °C; ρ in kg m^{-3}; h in kJ kg^{-1}; C_P in kJ kg^{-1} K^{-1}; S in kJ kg^{-1} K^{-1}

Supercritical steam properties

	T	400	425	450	500	550	600	650
22,500 kPa	ρ	127.19	105.14	92.895	78.314	69.244	62.743	57.720
	h	2713.1	2883.4	3008.2	3204.3	3368.0	3516.4	3656.5
	C_P	8.5750	5.6282	4.4903	3.5103	3.0887	2.8699	2.7464
	S	5.3655	5.6142	5.7899	6.0524	6.2578	6.4329	6.5890
25,000 kPa	ρ	166.54	126.81	108.98	89.744	78.517	70.720	64.810
	h	2578.6	2805.0	2950.6	3165.9	3339.2	3493.5	3637.7
	C_P	13.031	6.8145	5.0832	3.7639	3.2338	2.9653	2.8145
	S	5.1400	5.4707	5.6759	5.9642	6.1816	6.3637	6.5242
27,500 kPa	ρ	237.85	153.80	127.30	101.97	88.187	78.929	72.046
	h	2383.6	2715.3	2888.3	3126.1	3309.8	3470.3	3618.6
	C_P	24.038	8.4955	5.8100	4.0452	3.3886	3.0650	2.8848
	S	4.8313	5.3166	5.5604	5.8789	6.1094	6.2987	6.4640
30,000 kPa	ρ	357.43	188.73	148.43	115.07	98.277	87.377	79.431
	h	2152.8	2611.8	2821.0	3084.7	3279.7	3446.7	3599.4
	C_P	25.534	10.892	6.6996	4.3560	3.5532	3.1688	2.9570
	S	4.4757	5.1473	5.4421	5.7956	6.0402	6.2373	6.4074
32,500 kPa	ρ	433.81	234.80	173.03	129.14	108.81	96.067	86.966
	h	2041.6	2494.1	2748.4	3042.0	3249.1	3423.0	3580.1
	C_P	15.902	13.854	7.7685	4.6968	3.7272	3.2766	3.0312
	S	4.3012	4.9616	5.3201	5.7138	5.9737	6.1788	6.3539
35,000 kPa	ρ	474.97	291.21	201.73	144.25	119.79	105.01	94.649
	h	1988.6	2373.4	2671.0	2997.9	3218.0	3398.9	3560.7
	C_P	11.675	15.627	8.9775	5.0663	3.9101	3.3879	3.1071
	S	4.2143	4.7751	5.1945	5.6331	5.9092	6.1228	6.3030
37,500 kPa	ρ	502.47	347.40	234.69	160.47	131.24	114.19	102.48
	h	1955.4	2272.4	2590.7	2952.6	3186.4	3374.7	3541.2
	C_P	9.8084	14.826	10.150	5.4603	4.1010	3.5026	3.1844
	S	4.1574	4.6192	5.0677	5.5533	5.8466	6.0690	6.2544
40,000 kPa	ρ	523.34	394.09	270.89	177.84	143.17	123.62	110.46
	h	1931.4	2199.0	2511.8	2906.5	3154.4	3350.4	3521.6
	C_P	8.7328	12.941	10.957	5.8701	4.2985	3.6199	3.2631
	S	4.1145	4.5044	4.9448	5.4744	5.7857	6.0170	6.2078
45,000 kPa	ρ	554.49	457.22	343.00	215.84	168.43	143.22	126.84
	h	1897.7	2111.1	2377.6	2813.2	3089.9	3301.5	3482.5
	C_P	7.4943	9.7432	10.837	6.6716	4.7046	3.8603	3.4231
	S	4.0507	4.3617	4.7367	5.3207	5.6681	5.9179	6.1196
50,000 kPa	ρ	577.79	497.70	402.04	257.07	195.41	163.72	143.74
	h	1874.4	2060.7	2284.7	2722.6	3025.3	3252.5	3443.4
	C_P	6.7899	8.1964	9.5730	7.2889	5.1048	4.1028	3.5845
	S	4.0029	4.2746	4.5896	5.1762	5.5563	5.8245	6.0373

T in °C; ρ in kg m^{-3}; h in kJ kg^{-1}; C_P in kJ kg^{-1} K^{-1}; S in kJ kg^{-1} K^{-1}

Supercritical steam properties (cont.)

	T	700	750	800	850	900	950	1000
	ρ	53.653	50.254	47.348	44.820	42.592	40.607	38.822
22,500 kPa	h	3791.9	3924.5	4055.6	4186.0	4316.2	4446.6	4577.4
	C_P	2.6747	2.6339	2.6126	2.6044	2.6050	2.6117	2.6227
	S	6.7318	6.8647	6.9898	7.1085	7.2220	7.3308	7.4356
	ρ	60.084	56.171	52.848	49.973	47.449	45.207	43.196
25,000 kPa	h	3776.0	3910.9	4043.8	4175.6	4307.1	4438.5	4570.2
	C_P	2.7260	2.6741	2.6451	2.6311	2.6274	2.6308	2.6392
	S	6.6702	6.8054	6.9322	7.0523	7.1668	7.2765	7.3820
	ρ	66.614	62.156	58.396	55.159	52.328	49.822	47.581
27,500 kPa	h	3760.0	3897.2	4031.9	4165.2	4297.9	4430.4	4563.0
	C_P	2.7786	2.7150	2.6779	2.6581	2.6500	2.6500	2.6557
	S	6.6131	6.7506	6.8792	7.0007	7.1162	7.2268	7.3331
	ρ	73.242	68.208	63.990	60.377	57.230	54.453	51.976
30,000 kPa	h	3743.9	3883.4	4020.0	4154.9	4288.8	4422.3	4555.8
	C_P	2.8322	2.7565	2.7111	2.6853	2.6727	2.6693	2.6722
	S	6.5598	6.6997	6.8300	6.9529	7.0695	7.1810	7.2880
	ρ	79.966	74.324	69.628	65.626	62.153	59.098	56.379
32,500 kPa	h	3727.7	3869.7	4008.1	4144.5	4279.7	4414.2	4548.6
	C_P	2.8867	2.7985	2.7446	2.7127	2.6955	2.6886	2.6888
	S	6.5097	6.6519	6.7841	6.9083	7.0260	7.1384	7.2461
	ρ	86.786	80.505	75.310	70.904	67.097	63.757	60.792
35,000 kPa	h	3711.6	3855.9	3996.3	4134.2	4270.6	4406.2	4541.5
	C_P	2.9421	2.8410	2.7783	2.7402	2.7184	2.7079	2.7054
	S	6.4622	6.6069	6.7409	6.8665	6.9853	7.0985	7.2069
	ρ	93.701	86.748	81.034	76.212	72.059	68.428	65.212
37,500 kPa	h	3695.3	3842.2	3984.4	4123.9	4261.5	4398.2	4534.4
	C_P	2.9983	2.8839	2.8123	2.7677	2.7413	2.7273	2.7219
	S	6.4171	6.5643	6.7000	6.8270	6.9469	7.0610	7.1701
	ρ	100.71	93.052	86.799	81.546	77.040	73.111	69.640
40,000 kPa	h	3679.1	3828.4	3972.6	4113.6	4252.5	4390.2	4527.3
	C_P	3.0551	2.9271	2.8463	2.7954	2.7642	2.7466	2.7385
	S	6.3740	6.5236	6.6612	6.7896	6.9106	7.0256	7.1355
	ρ	114.99	105.83	98.443	92.292	87.052	82.508	78.513
45,000 kPa	h	3646.8	3801.1	3949.1	4093.1	4234.6	4374.4	4513.3
	C_P	3.1701	3.0141	2.9147	2.8507	2.8100	2.7851	2.7714
	S	6.2930	6.4477	6.5889	6.7201	6.8433	6.9601	7.0713
	ρ	129.59	118.82	110.22	103.13	97.121	91.941	87.405
50,000 kPa	h	3614.6	3773.9	3925.8	4072.9	4216.8	4358.7	4499.4
	C_P	3.2857	3.1013	2.9831	2.9059	2.8556	2.8235	2.8041
	S	6.2178	6.3775	6.5225	6.6565	6.7819	6.9004	7.0131

T in °C; ρ in kg m^{-3}; h in kJ kg^{-1}; C_P in kJ kg^{-1} K^{-1}; S in kJ kg^{-1} K^{-1}

19

Supercritical steam properties (cont.)

	T	400	425	450	500	550	600	650
	ρ	596.57	527.33	446.02	298.98	223.75	185.01	161.11
55,000 kPa	h	1856.9	2026.7	2223.1	2640.5	2962.2	3204.1	3404.8
	C_P	6.3300	7.3095	8.3631	7.5664	5.4662	4.3394	3.7444
	S	3.9643	4.2120	4.4882	5.0466	5.4505	5.7362	5.9598
	ρ	612.42	550.68	479.51	338.73	252.83	206.91	178.87
60,000 kPa	h	1843.2	2001.8	2180.2	2570.3	2901.9	3156.8	3366.7
	C_P	6.0022	6.7255	7.5265	7.5206	5.7539	4.5607	3.8996
	S	3.9317	4.1630	4.4140	4.9356	5.3517	5.6527	5.8867
	ρ	626.19	569.96	506.26	374.56	281.91	229.19	196.91
65,000 kPa	h	1832.1	1982.6	2148.4	2512.7	2845.8	3111.0	3329.6
	C_P	5.7543	6.3121	6.9427	7.2817	5.9440	4.7571	4.0465
	S	3.9032	4.1227	4.3559	4.8429	5.2608	5.5740	5.8176
	ρ	638.41	586.43	528.42	405.97	310.20	251.58	215.13
70,000 kPa	h	1822.9	1967.3	2123.7	2466.1	2794.9	3067.4	3293.5
	C_P	5.5586	6.0041	6.5109	6.9675	6.0338	4.9206	4.1820
	S	3.8779	4.0883	4.3084	4.7660	5.1785	5.5002	5.7522
	ρ	649.43	600.84	547.31	433.26	337.12	273.78	233.39
75,000 kPa	h	1815.3	1954.7	2104.0	2428.3	2749.6	3026.3	3258.7
	C_P	5.3992	5.7649	6.1790	6.6511	6.0396	5.0458	4.3031
	S	3.8549	4.0582	4.2683	4.7018	5.1047	5.4313	5.6904
	ρ	659.49	613.68	563.74	457.04	362.30	295.53	251.56
80,000 kPa	h	1808.8	1944.2	2087.8	2397.4	2709.9	2988.1	3225.5
	C_P	5.2661	5.5730	5.9168	6.3665	5.9852	5.1317	4.4073
	S	3.8340	4.0314	4.2335	4.6473	5.0391	5.3674	5.6321
	ρ	668.75	625.27	578.30	477.95	385.57	316.59	269.52
85,000 kPa	h	1803.3	1935.3	2074.3	2371.8	2675.2	2952.9	3194.0
	C_P	5.1530	5.4149	5.7049	6.1205	5.8939	5.1810	4.4934
	S	3.8146	4.0072	4.2027	4.6004	4.9807	5.3084	5.5771
	ρ	677.35	635.87	591.38	496.53	406.95	336.78	287.15
90,000 kPa	h	1798.6	1927.8	2062.9	2350.3	2645.1	2920.7	3164.3
	C_P	5.0553	5.2820	5.5300	5.9094	5.7835	5.1992	4.5611
	S	3.7966	3.9850	4.1751	4.5592	4.9288	5.2540	5.5255
	ρ	685.38	645.63	603.26	513.20	426.56	356.01	304.34
95,000 kPa	h	1794.6	1921.3	2053.1	2331.9	2618.8	2891.5	3136.4
	C_P	4.9698	5.1682	5.3831	5.7271	5.6669	5.1928	4.6110
	S	3.7797	3.9645	4.1501	4.5227	4.8823	5.2040	5.4770
	ρ	692.93	654.70	614.16	528.28	444.55	374.21	321.02
100,000 kPa	h	1791.1	1915.7	2044.7	2316.2	2595.9	2865.1	3110.5
	C_P	4.8942	5.0696	5.2577	5.5688	5.5516	5.1682	4.6447
	S	3.7639	3.9455	4.1271	4.4900	4.8405	5.1581	5.4315

T in $^\circ$C; $\quad$ ρ in kg m^{-3}; $\quad$ h in kJ kg^{-1}; $\quad$ C_P in kJ kg^{-1} K^{-1}; $\quad$ S in kJ kg^{-1} K^{-1}

Supercritical steam properties (cont.)

	T	700	750	800	850	900	950	1000
	ρ	144.48	131.99	122.12	114.04	107.24	101.40	96.308
55,000 kPa	h	3582.7	3747.0	3902.7	4052.9	4199.3	4343.3	4485.7
	C_P	3.4005	3.1880	3.0511	2.9607	2.9008	2.8614	2.8365
	S	6.1475	6.3122	6.4609	6.5976	6.7252	6.8454	6.9595
	ρ	159.62	145.32	134.12	125.01	117.39	110.87	105.22
60,000 kPa	h	3551.3	3720.5	3880.0	4033.1	4182.0	4328.1	4472.2
	C_P	3.5133	3.2735	3.1181	3.0148	2.9454	2.8989	2.8685
	S	6.0814	6.2510	6.4033	6.5428	6.6725	6.7944	6.9099
	ρ	174.94	158.77	146.19	136.02	127.56	120.36	114.12
65,000 kPa	h	3520.4	3694.4	3857.6	4013.7	4165.0	4313.1	4458.9
	C_P	3.6224	3.3569	3.1838	3.0679	2.9893	2.9358	2.8999
	S	6.0190	6.1934	6.3492	6.4914	6.6232	6.7468	6.8636
	ρ	190.39	172.30	158.31	147.06	137.74	129.84	123.01
70,000 kPa	h	3490.3	3668.9	3835.7	3994.7	4148.3	4298.4	4445.9
	C_P	3.7264	3.4375	3.2478	3.1199	3.0323	2.9720	2.9309
	S	5.9599	6.1389	6.2981	6.4430	6.5768	6.7021	6.8203
	ρ	205.90	185.87	170.46	158.11	147.92	139.31	131.89
75,000 kPa	h	3461.0	3644.0	3814.2	3976.0	4132.0	4283.9	4433.1
	C_P	3.8240	3.5147	3.3096	3.1704	3.0742	3.0074	2.9611
	S	5.9039	6.0872	6.2498	6.3971	6.5330	6.6599	6.7794
	ρ	221.41	199.44	182.60	169.15	158.08	148.75	140.74
80,000 kPa	h	3432.7	3619.7	3793.3	3957.7	4115.9	4269.8	4420.5
	C_P	3.9138	3.5878	3.3690	3.2192	3.1150	3.0420	2.9907
	S	5.8507	6.0382	6.2038	6.3537	6.4915	6.6199	6.7407
	ρ	236.84	212.99	194.71	180.16	168.21	158.17	149.56
85,000 kPa	h	3405.4	3596.2	3772.9	3939.9	4100.3	4255.9	4408.2
	C_P	3.9950	3.6563	3.4257	3.2663	3.1545	3.0755	3.0196
	S	5.8003	5.9915	6.1601	6.3123	6.4520	6.5819	6.7040
	ρ	252.13	226.45	206.77	191.13	178.31	167.54	158.34
90,000 kPa	h	3379.3	3573.4	3753.0	3922.6	4085.0	4242.4	4396.2
	C_P	4.0669	3.7197	3.4792	3.3113	3.1926	3.1081	3.0477
	S	5.7524	5.9470	6.1184	6.2729	6.4144	6.5458	6.6690
	ρ	267.22	239.80	218.76	202.03	188.34	176.87	167.07
95,000 kPa	h	3354.4	3551.5	3733.8	3905.7	4070.1	4229.2	4384.5
	C_P	4.1289	3.7776	3.5295	3.3542	3.2292	3.1395	3.0749
	S	5.7070	5.9046	6.0786	6.2352	6.3784	6.5112	6.6357
	ρ	282.04	253.00	230.64	212.86	198.32	186.14	175.75
100,000 kPa	h	3330.7	3530.5	3715.3	3889.3	4055.6	4216.3	4373.0
	C_P	4.1810	3.8298	3.5764	3.3949	3.2643	3.1698	3.1013
	S	5.6639	5.8642	6.0406	6.1991	6.3440	6.4782	6.6038

T in °C; ρ in kg m^{-3}; h in kJ kg^{-1}; C_P in kJ kg^{-1} K^{-1}; S in kJ kg^{-1} K^{-1}

Water density at 101.325 kPa

T (°C)	ρ (kg m^{-3})	T (°C)	ρ (kg m^{-3})	T (°C)	ρ (kg m^{-3})
0.0	999.84	21.0	998.00	61.0	982.68
0.5	999.87	22.0	997.77	62.0	982.16
1.0	999.90	23.0	997.54	63.0	981.63
1.5	999.92	24.0	997.30	64.0	981.09
2.0	999.94	25.0	997.05	65.0	980.55
2.5	999.96	26.0	996.79	66.0	980.00
3.0	999.97	27.0	996.52	67.0	979.45
3.5	999.97	28.0	996.24	68.0	978.90
4.0	999.97	29.0	995.95	69.0	978.33
4.5	999.97	30.0	995.65	70.0	977.76
5.0	999.97	31.0	995.34	71.0	977.19
5.5	999.96	32.0	995.03	72.0	976.61
6.0	999.94	33.0	994.70	73.0	976.03
6.5	999.93	34.0	994.37	74.0	975.44
7.0	999.90	35.0	994.03	75.0	805.60
7.5	999.88	36.0	993.69	76.0	804.50
8.0	999.85	37.0	993.33	77.0	803.40
8.5	999.82	38.0	992.97	78.0	802.30
9.0	999.78	39.0	992.60	79.0	801.20
9.5	999.74	40.0	992.22	80.0	800.10
10.0	999.70	41.0	991.83	81.0	799.00
10.5	999.66	42.0	991.44	82.0	797.90
11.0	999.61	43.0	991.04	83.0	796.81
11.5	999.56	44.0	990.63	84.0	969.26
12.0	999.50	45.0	990.21	85.0	968.61
12.5	999.44	46.0	989.79	86.0	967.96
13.0	999.38	47.0	989.36	87.0	967.31
13.5	999.32	48.0	988.93	88.0	966.64
14.0	999.25	49.0	988.48	89.0	965.98
14.5	999.18	50.0	988.04	90.0	965.31
15.0	999.10	51.0	987.58	91.0	964.63
15.5	999.03	52.0	987.12	92.0	963.96
16.0	998.95	53.0	986.65	93.0	963.27
16.5	998.86	54.0	986.17	94.0	962.58
17.0	998.78	55.0	985.69	95.0	961.89
17.5	998.69	56.0	985.21	96.0	961.19
18.0	998.60	57.0	984.71	97.0	960.49
18.5	998.50	58.0	984.21	98.0	959.78
19.0	998.41	59.0	983.71	99.0	959.07
19.5	998.31	60.0	983.20	100.0	958.37
20.0	998.21				

Properties of saturated water and steam

T ($°C$)	P (kPa)	C_{PL} (kJ kg^{-1} K^{-1})	C_{PV} (kJ kg^{-1} K^{-1})	μ_L (μPa)	μ_V (μPa)	k_L (mW m^{-1} K^{-1})	k_V (mW m^{-1} K^{-1})
0.01	0.6117	4.220	1.884	1791	9.22	561.0	17.07
5	0.8726	4.206	1.889	1518	9.34	570.5	17.34
10	1.2282	4.196	1.895	1306	9.46	580.0	17.62
15	1.7058	4.189	1.900	1138	9.59	589.3	17.92
20	2.3393	4.184	1.906	1002	9.73	598.4	18.23
25	3.1699	4.182	1.912	890.1	9.87	607.2	18.55
30	4.2470	4.180	1.918	797.4	10.01	615.5	18.89
35	5.6290	4.180	1.925	719.3	10.16	623.3	19.24
40	7.3849	4.180	1.931	653.0	10.31	630.6	19.60
45	9.5950	4.180	1.939	596.1	10.46	637.3	19.97
50	12.352	4.182	1.947	546.8	10.62	643.6	20.36
55	15.762	4.183	1.955	504.0	10.77	649.2	20.77
60	19.946	4.185	1.965	466.4	10.93	654.3	21.19
65	25.042	4.188	1.975	433.2	11.10	659.0	21.62
70	31.201	4.190	1.986	403.9	11.26	663.1	22.07
75	38.595	4.193	1.999	377.7	11.43	666.8	22.53
80	47.415	4.197	2.012	354.3	11.59	670.0	23.01
85	57.867	4.201	2.027	333.3	11.76	672.8	23.51
90	70.182	4.205	2.043	314.4	11.93	675.3	24.02
95	84.609	4.210	2.061	297.3	12.10	677.3	24.55
100	101.33	4.216	2.080	281.7	12.27	679.1	25.10
110	143.38	4.228	2.124	254.7	12.61	681.7	26.24
120	198.67	4.244	2.177	232.1	12.96	683.2	27.47
130	270.28	4.262	2.239	212.9	13.30	683.7	28.76
140	361.54	4.283	2.311	196.5	13.65	683.3	30.14
150	476.16	4.307	2.394	182.5	13.99	682.0	31.60
160	618.23	4.335	2.488	170.2	14.34	680.0	33.13
170	792.19	4.368	2.594	159.6	14.68	677.0	34.75
180	1002.8	4.405	2.713	150.1	15.03	673.3	36.45
190	1255.2	4.447	2.844	141.8	15.37	668.8	38.24
200	1554.9	4.496	2.990	134.3	15.71	663.3	40.11
210	1907.7	4.551	3.150	127.6	16.06	657.0	42.09
220	2319.6	4.615	3.329	121.5	16.41	649.6	44.17
230	2797.1	4.688	3.529	116.0	16.76	641.3	46.38
240	3346.9	4.772	3.754	110.9	17.13	631.8	48.73
250	3976.2	4.870	4.011	106.1	17.49	621.2	51.26
260	4692.3	4.986	4.308	101.7	17.88	609.2	54.03
270	5503.0	5.123	4.656	97.50	18.28	595.9	57.11
280	6416.6	5.289	5.073	93.51	18.70	581.1	60.61
290	7441.8	5.493	5.582	89.66	19.15	565.0	64.71
300	8587.9	5.750	6.220	85.90	19.65	547.4	69.65
310	9865.1	6.085	7.045	82.17	20.21	528.7	75.84
320	11,284	6.537	8.159	78.41	20.85	509.2	83.91
330	12,858	7.186	9.753	74.54	21.61	489.1	94.94
340	14,601	8.208	12.236	70.43	22.55	468.5	110.9
350	16,529	10.116	16.692	65.88	23.82	447.4	135.9

Correlations for the properties of water and steam

The correlations presented on this page allow various properties of water and steam to be **estimated**. They do not calculate exact values for the properties.

Vapour pressure as a function of temperature

$$\ln\left(\frac{p_v}{p_c}\right) = \frac{T_c}{T}(a_1\theta + a_2\theta^{3/2} + a_3\theta^3 + a_4\theta^{7/2} + a_5\theta^4 + a_6\theta^{15/2})$$

where, p_v is expressed in MPa, T is expressed in kelvin and θ is defined below. 273.16 K < T < 645 K

$$T_c = 647.096 \text{ K} \quad \text{and} \quad p_c = 22.064 \text{ MPa}$$

$a_1 = -7.859\,517\,83$ $a_2 = 1.844\,082\,59$ $a_3 = -11.786\,649\,7$
$a_4 = 22.680\,741\,1$ $a_5 = -15.961\,871\,9$ $a_6 = 1.801\,225\,02$

Saturated liquid density as a function of temperature

$$\frac{\rho_l}{\rho_c} = 1 + b_1\theta^{1/3} + b_2\theta^{2/3} + b_3\theta^{5/3} + b_4\theta^{16/3} + b_5\theta^{43/3} + b_6\theta^{110/3}$$

where, ρ_l is expressed in kg m^{-3} and θ is defined below. 273.16 K < T < 645 K
$$T_c = 647.096 \text{ K and } \rho_c = 322 \text{ kg m}^{-3}$$

$b_1 = 1.992\,740\,64$ $b_2 = 1.099\,653\,42$ $b_3 = -0.510\,839\,303$
$b_4 = -1.754\,934\,79$ $b_5 = -45.517\,035\,2$ $b_6 = -6.746\,944\,50 \times 10^5$

Saturated vapour density as a function of temperature

$$\ln\left(\frac{\rho_g}{\rho_c}\right) = c_1\theta^{1/3} + c_2\theta^{2/3} + c_3\theta^{4/3} + c_4\theta^3 + c_5\theta^{37/6} + c_6\theta^{71/6}$$

where, ρ_g is expressed in kg m^{-3} and θ is defined below. 273.16 K < T < 645 K
$$T_c = 647.096 \text{ K and } \rho_c = 322 \text{ kg m}^{-3}$$

$c_1 = -2.031\,502\,40$ $c_2 = -2.683\,029\,40$ $c_3 = -5.386\,264\,92$
$c_4 = -17.299\,160\,5$ $c_5 = -44.758\,658\,1$ $c_6 = -63.920\,106\,3$

In the above correlations, $\theta = \left(1 - \dfrac{T}{T_c}\right)$

Data on pages 2 to 24 is based upon the equation of state released by the International Association for the Properties of Water and Steam, published by Wagner, W., and Pruß, A., 2002, *J. Phys. Chem. Ref. Data* **31(2)**: 387–535. Data on page 23 is also based upon that of the International Association for the Properties of Water and Steam, Revised Release on the IAPS Formulation 1985 for the Thermal Conductivity of Ordinary Water Substance, September 1998; and Revised Formulation on the IAPS Formulation 1985 for the Viscosity of Ordinary Water Substance, August 2003 (www.iapws.org).

Vapour pressures of selected compounds

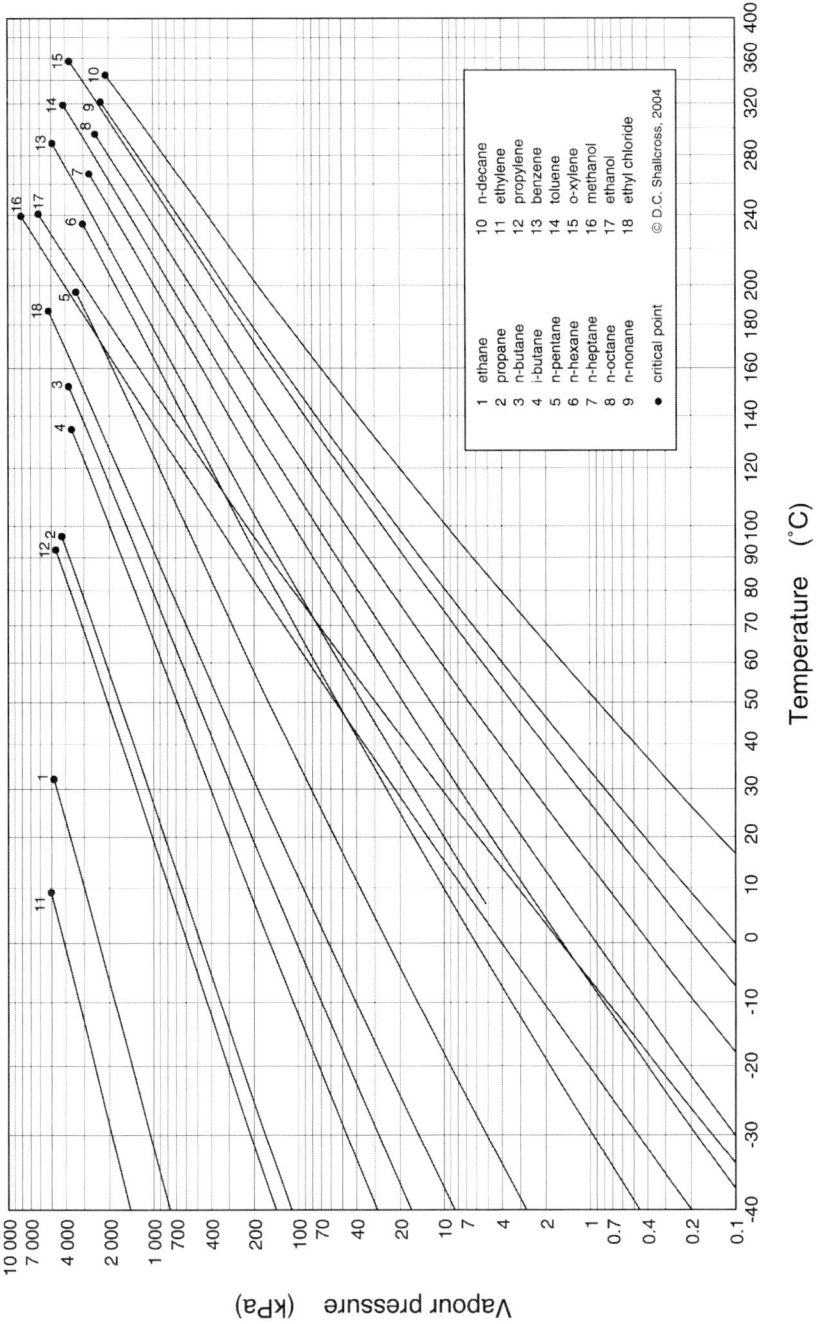

1	ethane	10	n-decane
2	propane	11	ethylene
3	n-butane	12	propylene
4	i-butane	13	benzene
5	n-pentane	14	toluene
6	n-hexane	15	o-xylene
7	n-heptane	16	methanol
8	n-octane	17	ethanol
9	n-nonane	18	ethyl chloride

• critical point

© D.C. Shallcross, 2004

25

Refrigerant 12 – dichlorodifluoromethane

T ($^\circ$C)	P (MPa)	ρ_L (kg m^{-3})	ρ_V (kg m^{-3})	h_f (kJ kg^{-1})	h_g (kJ kg^{-1})	Superheated enthalpies (kJ kg^{-1}) Degrees of superheat			
						25 $^\circ$C	50 $^\circ$C	75 $^\circ$C	100 $^\circ$C
−50.00	0.03912	1544	2.603	−44.82	129.40	143.03	157.26	172.07	187.43
−45.16	0.05000	1530	3.270	−40.64	131.67	145.51	159.91	174.86	190.36
−45.00	0.05041	1530	3.294	−40.49	131.74	145.59	160.00	174.96	190.46
−40.00	0.06415	1516	4.121	−36.14	134.09	148.16	162.74	177.86	193.49
−35.00	0.08070	1501	5.101	−31.75	136.43	150.72	165.49	180.76	196.52
−30.10	0.1000	1487	6.228	−27.42	138.71	153.23	168.18	183.60	199.49
−30.00	0.1004	1487	6.253	−27.33	138.76	153.28	168.23	183.66	199.55
−25.00	0.1237	1472	7.595	−22.88	141.08	155.84	170.97	186.55	202.59
−20.17	0.1500	1458	9.092	−18.54	143.31	158.29	173.61	189.35	205.51
−20.00	0.1510	1457	9.150	−18.39	143.39	158.38	173.71	189.45	205.61
−15.00	0.1827	1442	10.94	−13.86	145.69	160.92	176.43	192.33	208.63
−12.55	0.2000	1435	11.91	−11.62	146.81	162.15	177.77	193.74	210.11
−10.00	0.2193	1427	12.98	−9.28	147.96	163.44	179.15	195.21	211.65
−5.00	0.2612	1412	15.31	−4.67	150.22	165.94	181.85	198.07	214.65
0.00	0.3089	1396	17.95	0.00	152.44	168.43	184.54	200.93	217.64
5.00	0.3628	1379	20.93	4.71	154.64	170.89	187.21	203.76	220.62
8.13	0.4000	1369	22.98	7.69	155.99	172.42	188.87	205.53	222.47
10.00	0.4236	1363	24.28	9.48	156.79	173.32	189.85	206.58	223.58
15.00	0.4916	1346	28.03	14.30	158.91	175.73	192.48	209.37	226.52
20.00	0.5674	1329	32.23	19.18	160.98	178.10	195.07	212.15	229.44
22.00	0.6000	1322	34.05	21.14	161.80	179.04	196.10	213.25	230.60
25.00	0.6516	1311	36.92	24.11	163.00	180.44	197.64	214.89	232.34
30.00	0.7446	1293	42.15	29.11	164.96	182.73	200.17	217.61	235.21
32.76	0.8000	1282	45.30	31.89	166.02	183.98	201.55	219.10	236.78
35.00	0.8470	1274	47.98	34.17	166.85	184.98	202.67	220.30	238.05
40.00	0.9594	1254	54.47	39.29	168.67	187.18	205.12	222.96	240.87
41.70	1.000	1247	56.83	41.05	169.27	187.91	205.95	223.85	241.82
45.00	1.082	1234	61.69	44.49	170.40	189.32	207.54	225.57	243.65
49.40	1.200	1215	68.74	49.13	171.83	191.15	209.62	227.85	246.07
50.00	1.217	1212	69.75	49.76	172.02	191.40	209.90	228.15	246.40
55.00	1.363	1190	78.75	55.12	173.53	193.41	212.22	230.69	249.11
56.21	1.400	1184	81.10	56.44	173.88	193.89	212.78	231.30	249.77
60.00	1.521	1167	88.83	60.58	174.91	195.35	214.49	233.19	251.79
62.34	1.600	1155	93.98	63.17	175.51	196.23	215.53	234.34	253.03
65.00	1.693	1142	100.2	66.14	176.14	197.21	216.69	235.64	254.42
67.93	1.800	1127	107.5	69.46	176.77	198.25	217.96	237.05	255.95
70.00	1.879	1116	113.0	71.83	177.18	198.98	218.84	238.04	257.02
73.07	2.000	1099	121.7	75.40	177.71	200.02	220.12	239.49	258.59
75.00	2.079	1087	127.5	77.68	177.99	200.65	220.92	240.38	259.57
80.00	2.295	1057	144.3	83.71	178.54	202.22	222.92	242.68	262.07
84.33	2.500	1029	161.5	89.23	178.65	203.43	224.57	244.59	264.17
85.00	2.528	1024	163.8	89.97	178.75	203.67	224.86	244.91	264.52
90.00	2.779	987.6	186.9	96.56	178.52	205.00	226.71	247.08	266.92
94.12	3.000	954.1	209.7	102.31	177.88	205.99	228.18	248.82	268.86
95.00	3.049	946.4	215.1	103.58	177.67	206.18	228.48	249.19	269.26
100.00	3.340	898.5	251.3	111.26	175.88	207.22	230.15	251.22	271.55
105.00	3.654	839.1	302.0	120.08	172.41	208.08	231.72	253.17	273.76
110.00	3.994	746.6	397.3	131.91	164.02	208.75	233.17	255.03	275.90
110.08	4.000	744.2	400.0	132.17	163.76	208.76	233.19	255.06	275.93
111.80	4.125	558.0	558.0	136.0	136.0				

Enthalpy datum condition is saturated liquid at 0.0°C and 0.3089 MPa.
Tabulated data is based upon the equations of state presented by Stewart, R.B., Jacobsen, R.T., and Penoncello, S.G., 1986, ASHRAE Thermodynamics Properties of Refrigerants (ASHRAE, New York, USA).

Refrigerant 32 – difluoromethane

T (°C)	P (MPa)	ρ_L (kg m^{-3})	ρ_V (kg m^{-3})	h_f (kJ kg^{-1})	h_g (kJ kg^{-1})	Superheated enthalpies (kJ kg^{-1}) Degrees of superheat			
						25°C	50°C	75°C	100°C
−80.00	0.01863	1288	0.6123	−129.98	280.73	299.39	318.25	337.65	357.80
−78.98	0.02000	1286	0.6541	−128.40	281.33	300.06	318.97	338.42	358.61
−75.00	0.02618	1275	0.8414	−122.17	283.69	302.68	321.78	341.41	361.77
−70.00	0.03606	1262	1.135	−114.34	286.57	305.93	325.29	345.15	365.74
−65.00	0.04880	1249	1.506	−106.50	289.38	309.13	328.77	348.89	369.71
−64.58	0.05000	1248	1.541	−105.84	289.61	309.39	329.06	349.20	370.04
−60.00	0.06495	1236	1.969	−98.62	292.11	312.28	332.23	352.60	373.67
−55.00	0.08517	1222	2.538	−90.72	294.74	315.36	335.64	356.30	377.63
−51.91	0.1000	1214	2.951	−85.82	296.32	317.24	337.73	358.57	380.06
−50.00	0.1101	1208	3.232	−82.78	297.28	318.38	339.01	359.96	381.56
−45.00	0.1406	1194	4.067	−74.80	299.70	321.33	342.33	363.59	385.48
−40.00	0.1774	1180	5.065	−66.77	302.02	324.20	345.59	367.18	389.36
−37.32	0.2000	1172	5.674	−62.46	303.21	325.70	347.31	369.09	391.43
−35.00	0.2214	1166	6.248	−58.69	304.21	326.98	348.79	370.73	393.22
−30.00	0.2734	1151	7.639	−50.55	306.27	329.67	351.92	374.23	397.04
−25.00	0.3346	1136	9.266	−42.34	308.20	332.26	354.97	377.66	400.81
−20.38	0.4000	1122	11.00	−34.69	309.84	334.55	357.73	380.79	404.26
−20.00	0.4058	1121	11.16	−34.06	309.97	334.74	357.95	381.04	404.54
−15.00	0.4881	1105	13.35	−25.69	311.58	337.10	360.83	384.35	408.22
−10.00	0.5826	1089	15.87	−17.24	313.02	339.34	363.63	387.59	411.83
−9.15	0.6000	1086	16.33	−15.79	313.25	339.71	364.09	388.13	412.44
−5.00	0.6906	1072	18.77	−8.67	314.26	341.44	366.32	390.74	415.39
−0.51	0.8000	1057	21.73	−0.88	315.21	343.21	368.64	393.51	418.53
0.00	0.8131	1055	22.09	0.00	315.30	343.41	368.90	393.82	418.88
5.00	0.9514	1038	25.89	8.80	316.11	345.22	371.37	396.80	422.30
6.62	1.000	1032	27.24	11.69	316.32	345.77	372.15	397.75	423.39
10.00	1.107	1020	30.23	17.74	316.66	346.87	373.72	399.69	425.64
15.00	1.281	1001	35.19	26.84	316.94	348.34	375.95	402.49	428.90
20.00	1.475	981.4	40.86	36.12	316.90	349.64	378.04	405.17	432.08
25.00	1.690	961.0	47.34	45.60	316.51	350.73	379.99	407.75	435.17
30.00	1.928	939.6	54.78	55.32	315.73	351.62	381.79	410.22	438.17
31.43	2.000	933.3	57.10	58.15	315.42	351.83	382.28	410.90	439.01
35.00	2.190	917.1	63.34	65.30	314.48	352.28	383.44	412.56	441.07
40.00	2.478	893.0	73.27	75.61	312.71	352.70	384.92	414.78	443.88
45.00	2.795	867.3	84.86	86.31	310.29	352.87	386.23	416.88	446.58
48.02	3.000	850.7	92.83	92.98	308.48	352.84	386.94	418.07	448.15
50.00	3.141	839.3	98.55	97.49	307.10	352.77	387.37	418.83	449.17
55.00	3.520	808.3	115.0	109.29	302.93	352.37	388.31	420.65	451.64
60.00	3.933	773.3	135.2	121.93	297.44	351.65	389.06	422.32	454.00
60.77	4.000	767.5	138.8	123.97	296.45	351.51	389.16	422.56	454.35
65.00	4.384	732.3	161.1	135.80	290.05	350.59	389.60	423.82	
70.00	4.877	680.9	196.7	151.73	279.52	349.16	389.90	425.16	
71.18	5.000	666.3	207.5	156.00	276.27	348.76	389.93	425.44	
75.00	5.417	605.9	255.6	172.39	261.72	347.28	389.92	426.28	
78.11	5.783	424.0	424.0	214.15	214.15				

Enthalpy datum condition is saturated liquid at 0.0°C and 0.8131 MPa.
Tabulated data based upon the equation of state of Tillner-Roth, R., and Yokozeki, A., 1997, *J. Phys. Chem. Ref. Data*, **26(6):** 1273–1328.

Refrigerant 123 – 2,2-dichloro-1,1,1-trifluoroethane

T (°C)	P (MPa)	ρ_L (kg m^{-3})	ρ_V (kg m^{-3})	h_f (kJ kg^{-1})	h_g (kJ kg^{-1})	Superheated enthalpies (kJ kg^{-1}) Degrees of superheat			
						25°C	50°C	75°C	100°C
−50.00	0.001768	1643	0.1461	−48.19	152.21	166.89	182.45	198.84	216.00
−45.00	0.002538	1631	0.2052	−43.48	155.03	169.91	185.64	202.19	219.50
−40.00	0.003575	1620	0.2831	−38.75	157.88	172.95	188.86	205.57	223.03
−35.00	0.004951	1609	0.3843	−34.00	160.75	176.01	192.10	208.98	226.59
−30.00	0.006748	1597	0.5136	−29.22	163.65	179.10	195.37	212.40	230.17
−25.00	0.009061	1585	0.6767	−24.42	166.58	182.22	198.66	215.85	233.76
−20.00	0.01200	1574	0.8800	−19.59	169.52	185.35	201.97	219.32	237.38
−15.00	0.01568	1562	1.130	−14.73	172.48	188.51	205.30	222.81	241.02
−10.00	0.02025	1550	1.435	−9.85	175.45	191.68	208.64	226.32	244.67
−5.00	0.02585	1538	1.803	−4.94	178.44	194.86	212.00	229.84	248.34
0.00	0.03265	1526	2.242	0.00	181.44	198.06	215.38	233.38	252.03
5.00	0.04082	1514	2.762	4.97	184.44	201.27	218.77	236.93	255.72
9.73	0.05000	1502	3.339	9.70	187.30	204.31	221.98	240.30	259.23
10.00	0.05057	1502	3.374	9.97	187.46	204.49	222.17	240.49	259.43
15.00	0.06209	1489	4.089	14.99	190.48	207.71	225.57	244.06	263.15
20.00	0.07561	1477	4.917	20.05	193.50	210.94	228.99	247.64	266.88
25.00	0.09136	1464	5.872	25.14	196.51	214.17	232.40	251.22	270.61
27.46	0.1000	1458	6.392	27.65	198.00	215.76	234.09	252.98	272.45
30.00	0.1096	1451	6.966	30.26	199.53	217.40	235.82	254.81	274.34
35.00	0.1305	1438	8.213	35.41	202.54	220.63	239.25	258.40	278.09
40.00	0.1545	1425	9.629	40.59	205.54	223.86	242.67	261.99	281.83
45.00	0.1817	1411	11.23	45.81	208.53	227.08	246.08	265.58	285.57
48.05	0.2000	1403	12.30	49.00	210.34	229.04	248.16	267.76	287.85
50.00	0.2125	1398	13.03	51.06	211.50	230.29	249.49	269.16	289.31
55.00	0.2471	1384	15.05	56.34	214.46	233.49	252.90	272.74	293.05
60.00	0.2859	1370	17.31	61.67	217.40	236.68	256.30	276.32	296.79
65.00	0.3292	1356	19.83	67.03	220.32	239.86	259.68	279.89	300.52
70.00	0.3772	1341	22.63	72.43	223.21	243.01	263.06	283.45	304.24
72.20	0.4000	1335	23.96	74.82	224.47	244.40	264.54	285.01	305.88
75.00	0.4304	1326	25.74	77.87	226.07	246.15	266.42	287.00	307.95
80.00	0.4891	1311	29.19	83.35	228.89	249.27	269.76	290.54	311.66
85.00	0.5536	1296	33.00	88.88	231.68	252.36	273.09	294.06	315.35
88.34	0.6000	1285	35.76	92.59	233.52	254.40	275.30	296.40	317.81
90.00	0.6242	1280	37.21	94.45	234.43	255.42	276.39	297.57	319.03
95.00	0.7014	1264	41.86	100.08	237.13	258.45	279.68	301.06	322.70
100.00	0.7855	1247	47.00	105.77	239.77	261.45	282.94	304.53	326.35
100.82	0.8000	1244	47.89	106.70	240.20	261.94	283.47	305.10	326.95
110.00	0.9760	1212	58.91	117.32	244.88	267.33	289.38	311.41	333.61
111.15	1.000	1208	60.45	118.66	245.45	268.00	290.11	312.20	334.44
120.00	1.199	1174	73.47	129.15	249.67	273.04	295.69	318.20	340.79
130.00	1.458	1134	91.38	141.32	254.07	278.53	301.87	324.88	347.88
131.50	1.500	1127	94.42	143.18	254.69	279.34	302.78	325.87	348.94
140.00	1.756	1088	113.7	153.92	257.94	283.78	307.88	331.44	354.87
147.25	2.000	1052	133.6	163.42	260.29	287.40	312.12	336.11	359.88
150.00	2.099	1037	142.2	167.11	261.06	288.73	313.70	337.86	361.76
160.00	2.490	975.7	180.2	181.13	263.02	293.33	319.32	344.12	
170.00	2.937	896.9	235.5	196.61	262.89	297.51	324.69	350.21	
171.30	3.000	884.4	245.0	198.81	262.61	298.02	325.37	350.99	
180.00	3.451	765.9	341.9	216.23	256.82	301.16	329.75		
183.68	3.662	550.0	550.0	237.39	237.39				

Enthalpy datum condition is saturated liquid at 0.0°C and 0.03265 MPa.
Tabulated data based upon the equation of state of Younglove, B.A., and McLinden, M.O., 1994, *J. Phys. Chem. Ref. Data*, **23(5)**: 731–779.

Refrigerant 123 – 2,2-dichloro-1,1,1-trifluoroethane

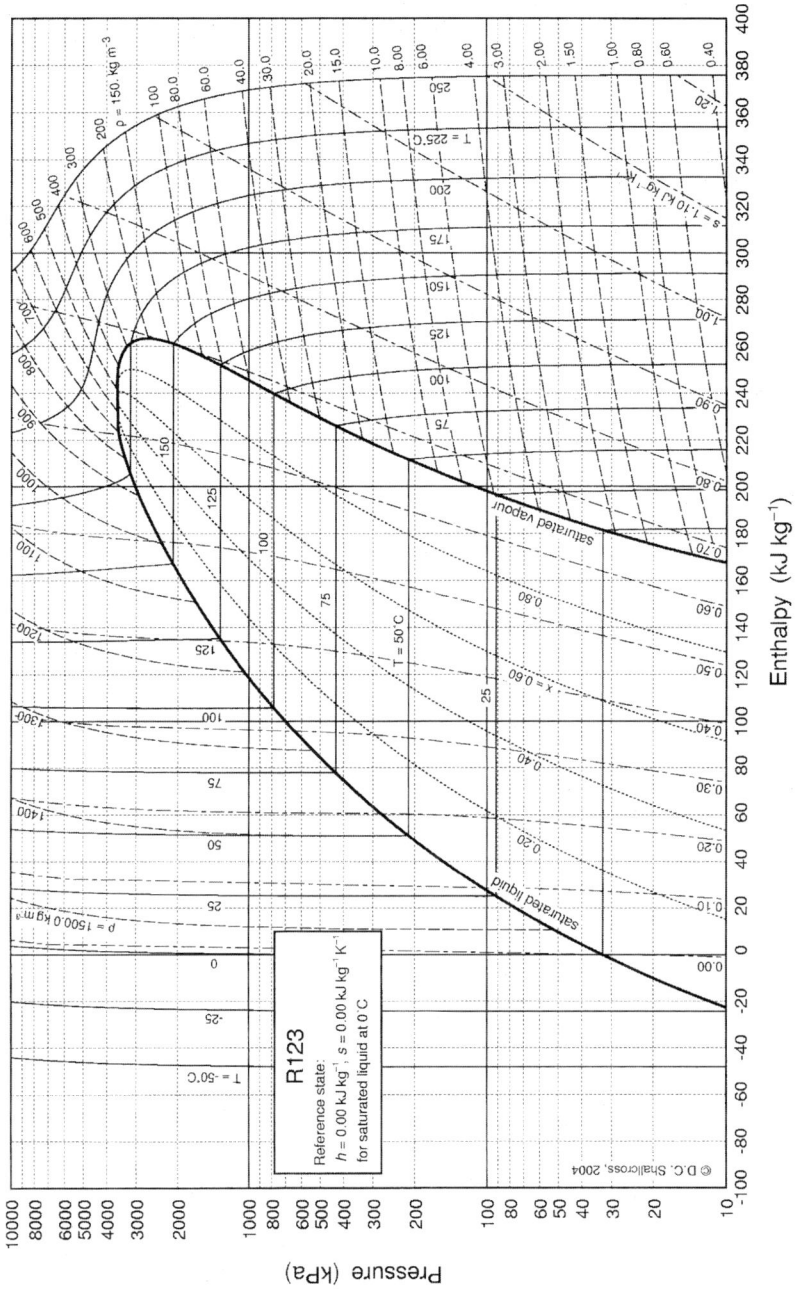

Presented data based upon the equation of state of Younglove, B.A., and McLinden, M.O., 1994, *J. Phys. Chem. Ref. Data*, **23(5)**: 731–779.

Refrigerant 134a – 1,1,1,2-tetrafluoroethane

T (°C)	P (MPa)	ρ_L (kg m^{-3})	ρ_V (kg m^{-3})	h_f (kJ kg^{-1})	h_g (kJ kg^{-1})	Superheated enthalpies (kJ kg^{-1}) Degrees of superheat			
						25°C	50°C	75°C	100°C
−50.00	0.02945	1446	1.650	−64.33	167.65	186.04	205.40	225.83	247.32
−45.00	0.03912	1432	2.152	−58.11	170.83	189.54	209.19	229.88	251.61
−40.45	0.05000	1419	2.708	−52.43	173.72	192.73	212.65	233.57	255.53
−40.00	0.05121	1418	2.770	−51.86	174.00	193.05	212.99	233.94	255.93
−35.00	0.06614	1403	3.521	−45.56	177.17	196.56	216.80	238.02	260.26
−30.00	0.08438	1388	4.426	−39.21	180.32	200.07	220.62	242.11	264.60
−26.36	0.1000	1378	5.193	−34.56	182.60	202.62	223.39	245.09	267.77
−25.00	0.1064	1373	5.506	−32.81	183.45	203.58	224.43	246.21	268.96
−20.00	0.1327	1358	6.785	−26.36	186.55	207.07	228.25	250.31	273.32
−17.13	0.1500	1349	7.617	−22.64	188.32	209.06	230.43	252.67	275.83
−15.00	0.1639	1343	8.287	−19.86	189.63	210.55	232.06	254.42	277.69
−10.08	0.2000	1327	10.01	−13.40	192.62	213.95	235.80	258.45	282.00
−10.00	0.2006	1327	10.04	−13.30	192.67	214.00	235.86	258.52	282.07
−5.00	0.2433	1311	12.08	−6.68	195.66	217.44	239.65	262.61	286.44
0.00	0.2928	1295	14.43	0.00	198.60	220.84	243.42	266.70	290.81
0.67	0.3000	1293	14.77	0.90	199.00	221.30	243.92	267.25	291.40
5.00	0.3497	1278	17.13	6.75	201.49	224.22	247.17	270.78	295.18
8.93	0.4000	1265	19.53	12.11	203.72	226.84	250.10	273.97	298.61
10.00	0.4146	1261	20.23	13.58	204.32	227.55	250.90	274.84	299.54
15.00	0.4884	1243	23.76	20.48	207.07	230.85	254.60	278.89	303.89
15.74	0.5000	1241	24.32	21.50	207.47	231.33	255.14	279.48	304.53
20.00	0.5717	1225	27.78	27.47	209.75	234.09	258.27	282.91	308.23
21.57	0.6000	1220	29.15	29.68	210.57	235.10	259.42	284.18	309.59
25.00	0.6654	1207	32.35	34.55	212.33	237.29	261.91	286.92	312.55
30.00	0.7702	1187	37.54	41.72	214.82	240.43	265.51	290.90	316.86
31.33	0.8000	1182	39.03	43.65	215.46	241.25	266.46	291.95	318.00
35.00	0.8870	1168	43.42	49.01	217.19	243.50	269.07	294.85	321.15
39.39	1.000	1149	49.22	55.50	219.16	246.14	272.15	298.29	324.89
40.00	1.017	1147	50.09	56.41	219.43	246.50	272.58	298.77	325.41
45.00	1.160	1125	57.66	63.94	221.52	249.43	276.05	302.65	329.66
46.32	1.200	1119	59.81	65.95	222.04	250.18	276.95	303.67	330.77
50.00	1.318	1102	66.27	71.62	223.44	252.27	279.46	306.50	333.87
52.42	1.400	1091	70.87	75.40	224.30	253.61	281.09	308.35	335.90
55.00	1.492	1078	76.10	79.47	225.15	255.02	282.81	310.31	338.06
57.91	1.600	1064	82.46	84.11	226.04	256.57	284.73	312.50	340.48
60.00	1.682	1053	87.38	87.51	226.63	257.67	286.10	314.07	342.22
62.90	1.800	1037	94.68	92.26	227.36	259.15	287.97	316.23	344.61
65.00	1.890	1026	100.4	95.76	227.82	260.20	289.32	317.79	346.34
67.48	2.000	1011	107.6	99.95	228.28	261.42	290.89	319.62	348.37
70.00	2.117	996.3	115.6	104.28	228.65	262.62	292.47	321.46	350.43
75.00	2.364	964.1	133.5	113.13	229.03	264.90	295.54	325.08	354.48
77.58	2.500	946.2	144.1	117.84	229.01	266.01	297.09	326.92	356.55
80.00	2.633	928.2	155.1	122.39	228.81	267.03	298.53	328.64	358.48
85.00	2.926	887.2	181.9	132.22	227.76	269.00	301.43	332.13	362.44
86.20	3.000	876.2	189.3	134.71	227.34	269.44	302.11	332.97	363.39
90.00	3.244	837.8	216.8	142.93	225.42	270.78	304.22	335.56	366.35
95.00	3.591	772.7	267.1	155.25	220.67	272.37	306.90	338.90	
100.00	3.972	651.2	373.0	173.30	207.69	273.69	309.42	342.14	
100.34	4.000	632.9	390.2	175.57	205.38	273.77	309.58	342.35	
101.06	4.059	512.0	512.0	189.63	189.63				

Enthalpy datum condition is saturated liquid at 0.0°C and 0.2928 MPa.

Tabulated data based upon the equation of state of Tillner-Roth, R., and Baehr, H.D., 1994, *J. Phys. Chem. Ref. Data*, **23(5):** 657–699.

Refrigerant 152a – 1,1-difluoroethane

T ($^\circ$C)	P (MPa)	ρ_L (kg m^{-3})	ρ_V (kg m^{-3})	h_f (kJ kg^{-1})	h_g (kJ kg^{-1})	Superheated enthalpies (kJ kg^{-1}) Degrees of superheat			
						25°C	50°C	75°C	100°C
−50.00	0.02743	1064	0.9936	−81.38	270.40	292.89	316.61	341.66	368.08
−45.00	0.03623	1054	1.289	−73.52	274.22	297.11	321.20	346.59	373.34
−40.00	0.04721	1044	1.651	−65.60	278.02	301.34	325.80	351.54	378.63
−38.88	0.05000	1042	1.742	−63.83	278.86	302.28	326.83	352.65	379.81
−35.00	0.06073	1034	2.089	−57.63	281.80	305.55	330.41	356.51	383.94
−30.00	0.07719	1024	2.615	−49.61	285.55	309.76	335.02	361.49	389.26
−25.00	0.09701	1013	3.241	−41.52	289.26	313.96	339.63	366.48	394.60
−24.32	0.1000	1012	3.334	−40.41	289.77	314.53	340.26	367.16	395.33
−20.00	0.1207	1003	3.979	−33.36	292.94	318.13	344.23	371.47	399.95
−15.00	0.1487	992.1	4.844	−25.14	296.57	322.28	348.83	376.46	405.31
−10.00	0.1815	981.3	5.852	−16.84	300.15	326.40	353.40	381.45	410.67
−7.49	0.2000	975.8	6.415	−12.65	301.92	328.45	355.69	383.95	413.36
−5.00	0.2198	970.3	7.017	−8.46	303.66	330.48	357.96	386.43	416.03
0.00	0.2640	959.1	8.359	0.00	307.11	334.53	362.50	391.39	421.39
3.61	0.3000	950.9	9.449	6.17	309.56	337.42	365.76	394.97	425.25
5.00	0.3148	947.7	9.897	8.55	310.49	338.53	367.01	396.34	426.74
10.00	0.3728	936.1	11.65	17.19	313.79	342.47	371.48	401.27	432.07
12.14	0.4000	931.0	12.48	20.92	315.17	344.15	373.39	403.38	434.36
15.00	0.4386	924.2	13.65	25.93	316.99	346.36	375.92	406.17	437.39
20.00	0.5129	912.0	15.91	34.77	320.09	350.19	380.31	411.05	442.70
25.00	0.5964	899.5	18.47	43.73	323.09	353.95	384.66	415.89	447.97
25.20	0.6000	899.0	18.58	44.09	323.21	354.10	384.83	416.09	448.19
30.00	0.6898	886.6	21.36	52.80	325.96	357.63	388.96	420.70	453.23
35.00	0.7939	873.4	24.61	62.01	328.70	361.23	393.20	425.47	458.45
35.28	0.8000	872.6	24.81	62.53	328.85	361.43	393.43	425.73	458.74
40.00	0.9093	859.7	28.28	71.35	331.28	364.75	397.38	430.19	463.64
43.61	1.000	849.5	31.21	78.18	333.05	367.22	400.35	433.57	467.36
45.00	1.037	845.5	32.41	80.84	333.70	368.16	401.49	434.87	468.80
50.00	1.177	830.8	37.06	90.50	335.93	371.47	405.54	439.49	473.91
50.76	1.200	828.5	37.82	91.99	336.26	371.97	406.15	440.19	474.69
55.00	1.332	815.4	42.30	100.34	337.95	374.67	409.51	444.06	478.98
57.07	1.400	808.9	44.67	104.48	338.72	375.96	411.13	445.94	481.07
60.00	1.501	799.4	48.22	110.38	339.72	377.74	413.39	448.57	484.01
62.74	1.600	790.2	51.80	115.98	340.58	379.37	415.49	451.01	486.75
65.00	1.685	782.5	54.93	120.64	341.21	380.68	417.19	453.01	488.99
67.90	1.800	772.2	59.24	126.72	341.93	382.32	419.36	455.56	491.85
70.00	1.886	764.6	62.57	131.16	342.37	383.47	420.90	457.39	493.91
72.65	2.000	754.7	67.05	136.85	342.83	384.89	422.82	459.68	496.50
75.00	2.105	745.6	71.31	141.98	343.14	386.10	424.51	461.70	498.78
80.00	2.342	725.2	81.40	153.15	343.43	388.57	428.01	465.92	503.59
83.11	2.500	711.6	88.51	160.30	343.32	390.00	430.13	468.51	506.55
85.00	2.600	703.0	93.19	164.74	343.13	390.84	431.40	470.07	508.33
90.00	2.878	678.5	107.2	176.87	342.06	392.90	434.66	474.13	513.01
92.07	3.000	667.5	113.8	182.09	341.33	393.70	435.98	475.79	514.93
95.00	3.179	650.9	124.2	189.71	339.95	394.75	437.80	478.10	517.61
100.00	3.505	618.5	145.8	203.59	336.28	396.34	440.80	481.96	522.13
105.00	3.858	578.1	175.2	219.25	329.96	397.65	443.62	485.71	526.56
106.89	4.000	558.9	190.1	226.00	326.35	398.06	444.65	487.09	528.20
110.00	4.243	517.4	224.3	239.22	317.31	398.60	446.25	489.30	530.86
113.26	4.517	368.0	368.0	277.55	277.55				

Enthalpy datum condition is saturated liquid at 0.0°C and 0.2640 MPa.

Tabulated data based upon the equation of state of Outcalt, S.L., and McLinden, M.O., 1996, *J. Phys. Chem. Ref. Data*, **25(2):** 605–636.

31

Refrigerant 717 – ammonia

T ($^\circ$C)	P (MPa)	ρ_L (kg m^{-3})	ρ_V (kg m^{-3})	h_f (kJ kg^{-1})	h_g (kJ kg^{-1})	Superheated enthalpies (kJ kg^{-1}) Degrees of superheat			
						25°C	50°C	75°C	100°C
−40.00	0.0717	689.9	0.6440	−180.4	1208.5	1262.5	1316.1	1369.8	1423.9
−35.00	0.0932	683.7	0.8223	−158.3	1216.2	1271.1	1325.3	1379.5	1434.0
−33.60	0.1000	682.0	0.8787	−152.1	1218.3	1273.5	1327.9	1382.2	1436.8
−30.00	0.1195	677.5	1.038	−136.1	1223.7	1279.6	1334.4	1389.1	1444.0
−25.00	0.1516	671.2	1.297	−113.8	1230.8	1287.8	1343.4	1398.5	1453.9
−20.00	0.1902	664.9	1.604	−91.3	1237.7	1295.8	1352.2	1407.9	1463.7
−18.86	0.2000	663.4	1.682	−86.2	1239.2	1297.6	1354.1	1410.0	1465.9
−15.00	0.2363	658.5	1.967	−68.7	1244.2	1303.6	1360.8	1417.1	1473.4
−10.00	0.2909	652.0	2.392	−46.0	1250.5	1311.1	1369.2	1426.1	1482.9
−9.24	0.3000	651.0	2.462	−42.5	1251.4	1312.3	1370.4	1427.5	1484.4
−5.00	0.3549	645.3	2.886	−23.1	1256.3	1318.4	1377.4	1435.0	1492.3
−1.89	0.4000	641.2	3.232	−8.7	1259.8	1322.8	1382.4	1440.4	1498.1
0.00	0.4296	638.6	3.458	0.0	1261.8	1325.4	1385.4	1443.7	1501.6
4.13	0.5000	633.0	3.994	19.2	1266.1	1331.0	1391.8	1450.7	1509.1
5.00	0.5159	631.8	4.115	23.2	1267.0	1332.2	1393.1	1452.2	1510.7
9.28	0.6000	625.8	4.753	43.2	1271.0	1337.7	1399.5	1459.3	1518.3
10.00	0.6152	624.8	4.868	46.6	1271.7	1338.6	1400.6	1460.5	1519.6
13.80	0.7000	619.4	5.510	64.5	1275.0	1343.2	1406.1	1466.6	1526.3
15.00	0.7286	617.7	5.727	70.2	1276.0	1344.7	1407.8	1468.5	1528.3
17.85	0.8000	613.6	6.266	83.7	1278.2	1348.0	1411.8	1473.0	1533.2
20.00	0.8575	610.4	6.701	93.9	1279.8	1350.4	1414.8	1476.3	1536.9
21.52	0.9000	608.2	7.023	101.2	1280.9	1352.1	1416.8	1478.7	1539.4
24.90	1.000	603.1	7.781	117.4	1283.1	1355.7	1421.3	1483.8	1545.0
25.00	1.003	603.0	7.805	117.9	1283.2	1355.8	1421.4	1483.9	1545.2
30.00	1.167	595.4	9.050	142.1	1286.0	1360.8	1427.8	1491.2	1553.3
30.94	1.200	593.9	9.302	146.6	1286.4	1361.7	1428.9	1492.6	1554.7
35.00	1.350	587.6	10.45	166.4	1288.2	1365.4	1433.8	1498.2	1561.1
36.26	1.400	585.6	10.83	172.7	1288.7	1366.5	1435.2	1500.0	1563.1
40.00	1.555	579.6	12.03	191.1	1289.9	1369.5	1439.4	1505.0	1568.7
41.03	1.600	577.9	12.38	196.2	1290.1	1370.3	1440.5	1506.3	1570.3
45.00	1.782	571.3	13.80	216.0	1290.9	1373.2	1444.7	1511.4	1576.1
45.37	1.800	570.7	13.94	217.8	1290.9	1373.5	1445.1	1511.9	1576.6
49.37	2.000	564.0	15.52	237.9	1291.2	1376.0	1449.0	1516.8	1582.3
50.00	2.033	562.9	15.78	241.1	1291.2	1376.4	1449.6	1517.5	1583.2
55.00	2.310	554.2	18.00	266.6	1290.7	1379.1	1454.1	1523.3	1590.0
58.17	2.500	548.5	19.55	282.9	1290.0	1380.5	1456.8	1526.8	1594.2
60.00	2.615	545.2	20.49	292.4	1289.4	1381.3	1458.2	1528.8	1596.6
65.00	2.948	535.9	23.28	318.7	1287.2	1382.9	1461.9	1533.9	1602.8
65.74	3.000	534.5	23.72	322.6	1286.8	1383.1	1462.4	1534.7	1603.8
70.00	3.312	526.2	26.41	345.3	1284.0	1383.9	1465.2	1538.7	1608.8
72.42	3.500	521.4	28.06	358.4	1282.1	1384.2	1466.6	1540.9	1611.6
75.00	3.709	516.2	29.92	372.5	1279.8	1384.4	1468.0	1543.1	1614.5
78.42	4.000	509.0	32.58	391.5	1276.1	1384.3	1469.6	1545.9	1618.3
80.00	4.141	505.6	33.89	400.3	1274.3	1384.2	1470.3	1547.2	1619.9
85.00	4.609	494.6	38.38	428.9	1267.4	1383.3	1472.1	1550.8	1625.0
88.90	5.000	485.5	42.29	451.7	1261.0	1382.1	1473.2	1553.4	1628.8
90.00	5.115	482.9	43.48	458.2	1259.0	1381.7	1473.5	1554.1	1629.8
95.00	5.663	470.4	49.31	488.6	1248.9	1379.3	1474.3	1557.0	1634.3
97.90	6.000	462.7	53.10	506.8	1242.0	1377.6	1474.6	1558.5	1636.7
100.00	6.254	457.0	56.06	520.3	1236.5	1376.2	1474.6	1559.5	1638.4

Enthalpy datum condition is saturated liquid at 0.0°C and 0.4296 MPa.

Tabulated data based upon the equation of state of Haar, L., and Gallagher, J.S., 1978, *J. Phys. Chem. Ref. Data*, **7(3):** 635–792.

Generalized compressibility factor chart

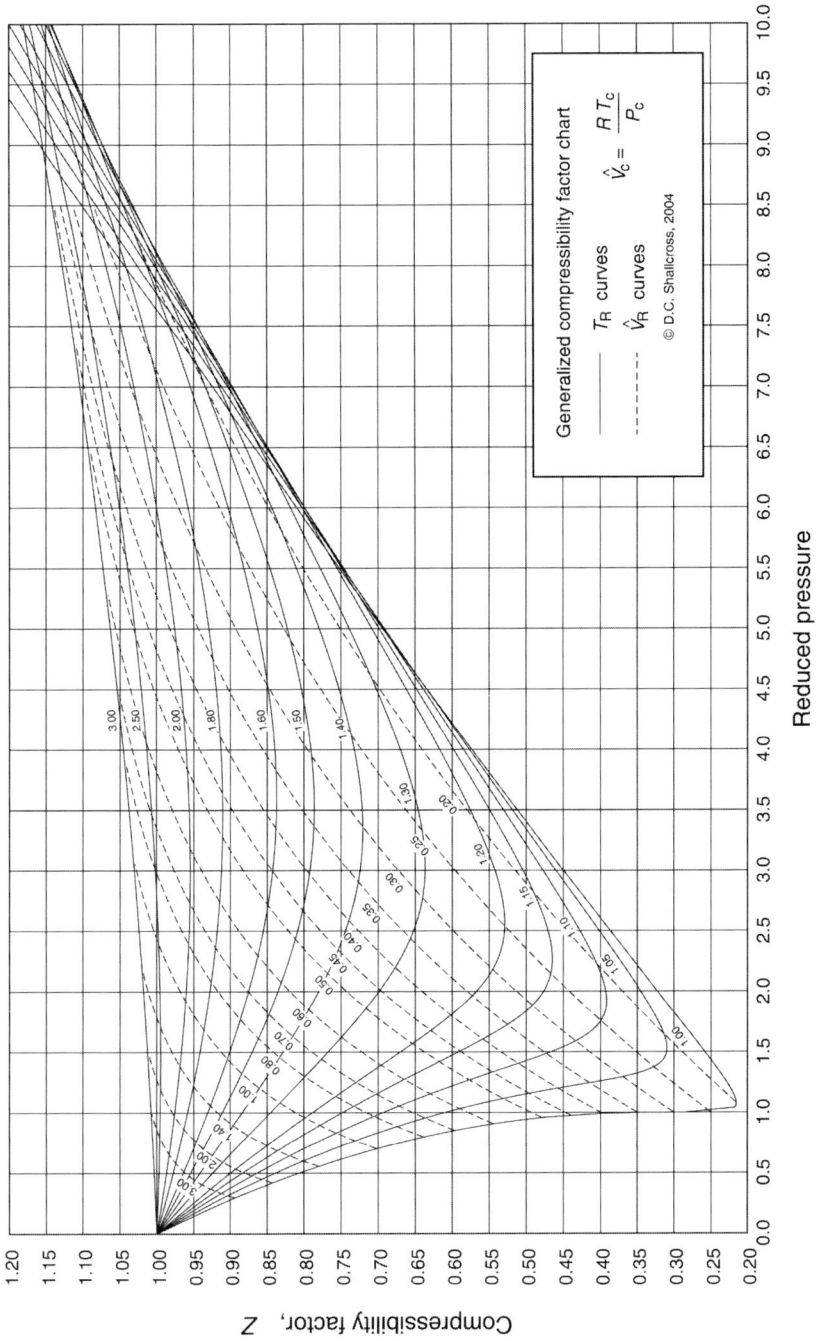

Generalized compressibility factor chart

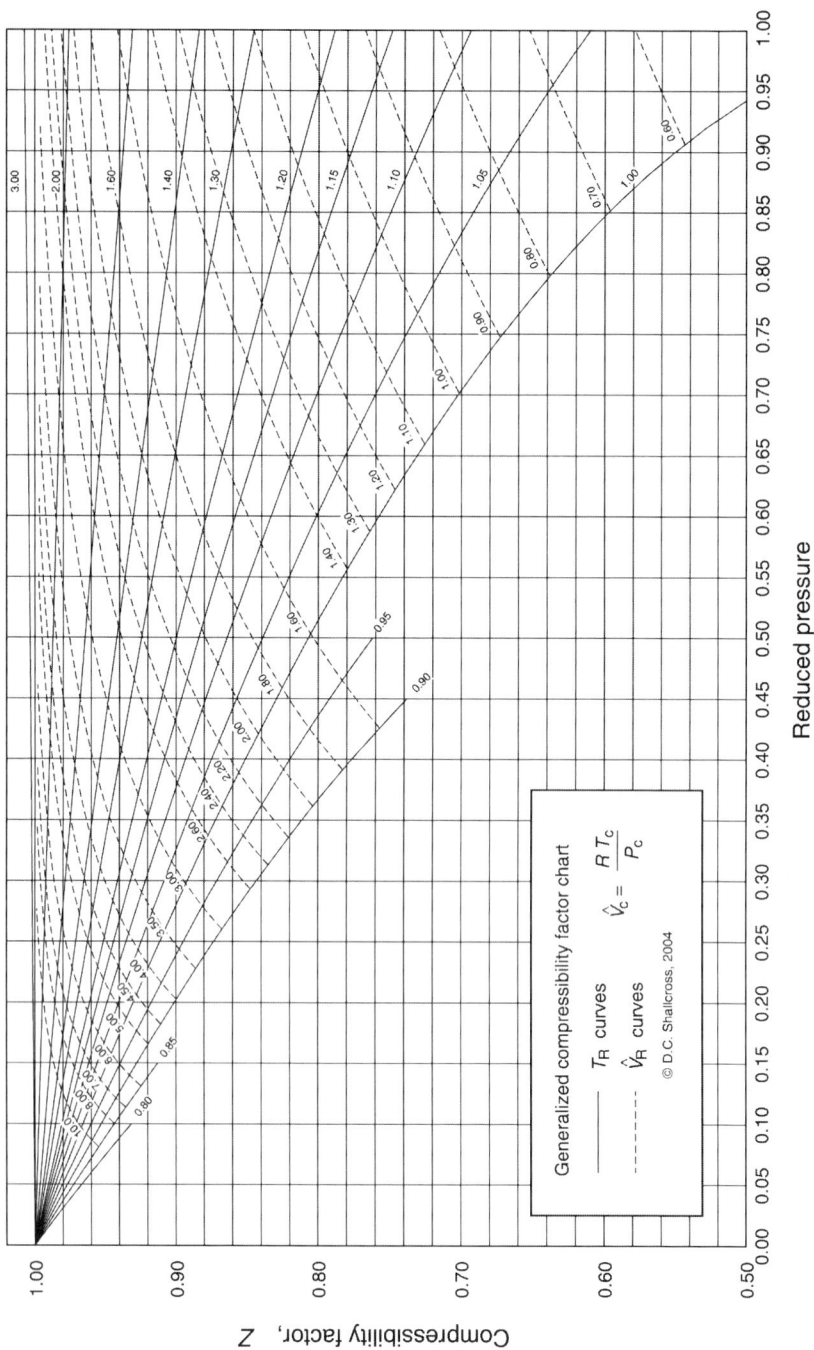

Generalized compressibility factor chart

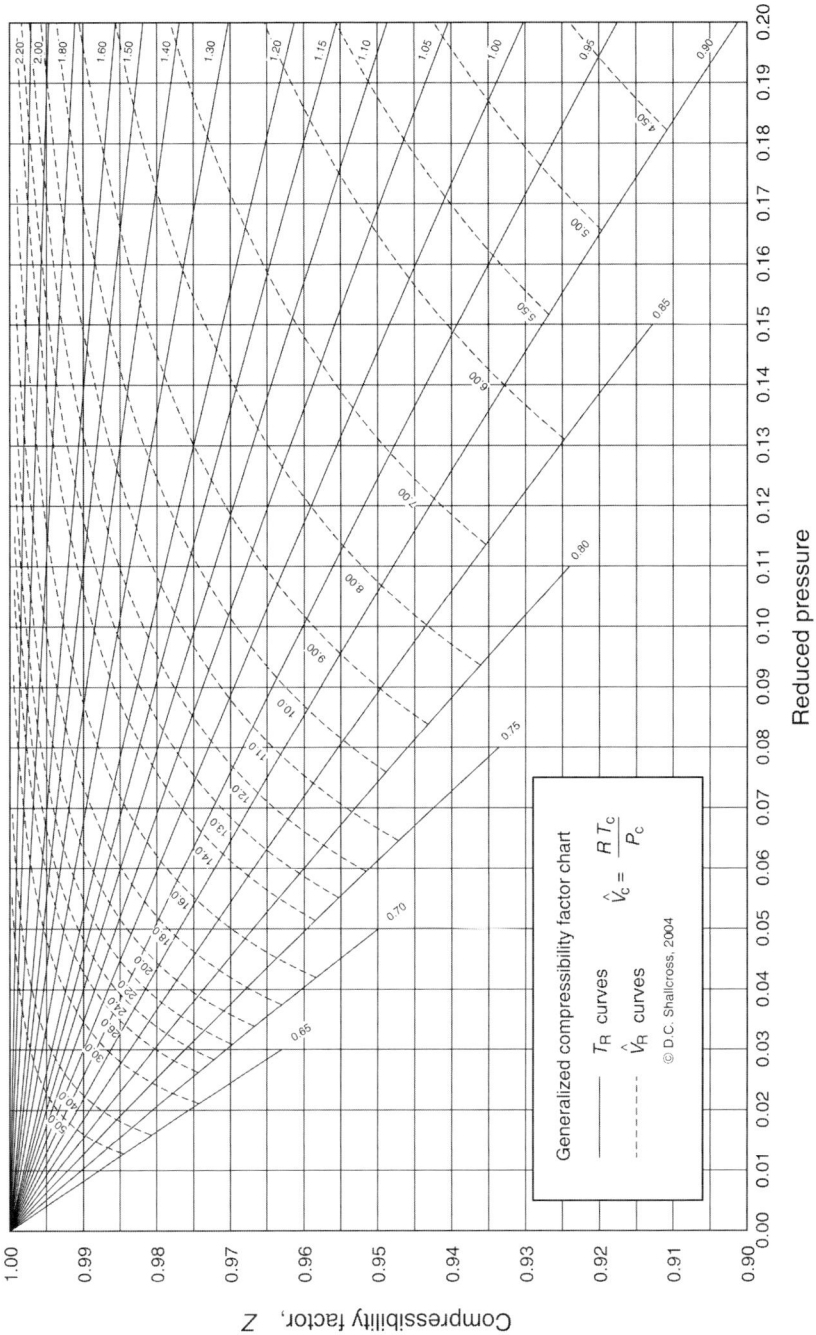

Generalized compressibility factor chart

— T_R curves

----- $\hat{V}_R$ curves

$\hat{V}_c = \dfrac{R\,T_c}{P_c}$

© D.C. Shallcross, 2004

Compressibility factor, Z

Reduced pressure

Methane, ethane and propane enthalpies

P (MPa)		0.020	0.050	0.10	0.20	0.50	1.00	2.00	5.00	10.0	20.0
	T_M	95.1	103.7	111.5	120.6	135.3	149.1	165.9	Saturation		
	T_E	159.1	172.4	184.3	198.2	220.4	241.1	266.0	temperatures		
	T_P	200.1	216.3	230.8	247.7	274.8	300.0	330.3	in K		
	h_M	374.2	358.2	344.8	330.5	311.7	300.7	300.7	Enthalpies at		
T_{sat}	h_E	165.5	149.4	135.6	120.3	98.6	83.0	73.9	saturation		
	h_P	100.9	81.4	64.0	43.8	12.6	13.4	37.0	temperatures		
	h_M	154.6	155.2	156.1	158.0	163.9	174.2	197.4	311.0	495.8	514.1
200 K	h_E	109.6	110.9	113.1	117.6						
	h_P										
	h_M	112.6	113.1	113.9	115.5	120.4	128.9	147.2	215.7	371.7	435.3
220 K	h_E	80.8	81.9	83.7	87.3						
	h_P	74.6	76.4								
	h_M	70.4	70.8	71.4	72.8	77.0	84.2	99.3	150.9	254.1	353.1
240 K	h_E	50.8	51.8	53.3	56.3	65.9					
	h_P	46.8	48.2	50.6							
	h_M	27.6	27.9	28.5	29.7	33.4	39.6	52.4	94.1	169.9	270.4
260 K	h_E	19.5	20.3	21.7	24.2	32.2	47.2				
	h_P	17.3	18.6	20.6	24.6						
	h_M	0.9	0.6	0.0	1.1	4.4	10.1	21.7	58.9	123.9	218.2
273 K	h_E	2.0	1.2	0.0	2.4	9.6	22.7	56.3			
	h_P	3.0	1.8	0.0	3.6						
	h_M	15.8	15.5	15.0	13.9	10.7	5.3	5.8	40.9	101.2	191.6
280 K	h_E	13.3	12.5	11.4	9.1	2.3	10.0	40.1			
	h_P	13.8	12.6	10.9	7.6	3.5					
	h_M	55.9	55.6	55.1	54.2	51.3	46.5	36.7	6.1	45.1	125.1
298 K	h_E	44.4	29.5	42.7	40.7	34.6	24.1	0.3	283.2	305.9	312.8
	h_P	43.5	42.5	41.0	38.1	28.8					
	h_M	60.0	59.7	59.2	58.3	55.5	50.7	41.0	10.8	39.5	118.5
300 K	h_E	47.6	29.5	45.9	43.9	37.9	27.5	3.7	274.5	299.7	307.5
	h_P	46.6	45.6	44.1	41.2	32.1					
	h_M	105.1	104.9	104.4	103.6	101.1	96.8	88.2	61.9	18.8	50.1
320 K	h_E	83.5	46.9	82.0	80.2	74.9	65.9	46.1	43.4	224.6	248.0
	h_P	81.2	80.3	78.9	76.4	68.7	53.9				

h_M, h_E and h_P are gas enthalpies of methane, ethane and propane respectively all in kJ kg^{-1}.
Enthalpy datum condition for each alkane is 0°C and 0.10 MPa.

Methane, ethane and propane enthalpies (cont.)

P (MPa)		0.020	0.050	0.10	0.20	0.50	1.00	2.00	5.00	10.0	20.0
340 K	h_M	151.3	151.0	150.7	149.9	147.7	143.9	136.2	113.0	75.4	14.9
	h_E	121.0	82.9	119.7	118.1	113.4	105.4	88.5	23.9	130.2	184.8
	h_P	117.5	116.7	115.5	113.3	106.6	94.3	62.2	204.8	209.2	208.6
360 K	h_M	198.6	198.4	198.1	197.4	195.4	191.9	185.1	164.4	131.3	77.8
	h_E	160.3	120.5	159.1	157.6	153.4	146.3	131.5	79.6	34.4	118.5
	h_P	155.7	155.0	153.9	151.9	146.0	135.4	110.3	135.2	149.4	153.8
380 K	h_M	247.2	247.0	246.7	246.1	244.3	241.2	235.0	216.6	187.2	139.5
	h_E	201.3	159.9	200.2	198.8	195.0	188.7	175.6	131.9	44.4	50.0
	h_P	195.7	195.0	194.0	192.2	186.9	177.8	156.9	3.7	84.4	96.6
400 K	h_M	297.1	297.0	296.7	296.2	294.5	291.7	286.2	269.6	243.3	200.7
	h_E	244.0	243.6	243.0	241.8	238.3	232.5	220.8	183.0	112.2	19.2
	h_P	237.5	236.8	235.9	234.3	229.5	221.4	203.6	124.6	12.3	36.7
420 K	h_M	348.5	348.3	348.1	347.6	346.1	343.6	338.5	323.6	300.1	261.8
	h_E	288.4	288.1	287.5	286.4	283.2	278.0	267.4	234.2	174.6	87.7
	h_P	281.0	280.4	279.5	278.0	273.7	266.5	251.0	191.2	68.4	25.6
440 K	h_M	401.3	401.1	400.9	400.4	399.1	396.8	392.2	378.8	357.7	323.2
	h_E	334.6	334.2	333.7	332.7	329.7	325.0	315.4	285.8	234.4	154.7
	h_P	326.2	325.7	324.9	323.5	319.6	313.0	299.3	250.4	150.1	89.6
460 K	h_M	455.5	455.4	455.2	454.8	453.6	451.5	447.4	435.2	416.2	385.2
	h_E	382.4	382.1	381.6	380.6	377.9	373.6	364.8	338.2	293.0	220.4
	h_P	373.1	372.6	371.9	370.6	367.0	361.1	348.8	307.2	225.4	155.0
480 K	h_M	511.3	511.2	511.0	510.6	509.5	507.6	503.9	492.9	475.7	447.8
	h_E	431.9	431.6	431.1	430.2	427.7	423.7	415.7	391.6	351.4	285.3
	h_P	421.7	421.2	420.6	419.4	416.1	410.7	399.6	363.3	295.3	221.6
500 K	h_M	568.5	568.4	568.3	567.9	566.9	565.2	561.8	551.9	536.4	511.3
	h_E	483.0	482.7	482.3	481.4	479.1	475.4	468.1	446.1	410.0	349.8
	h_P	471.9	471.5	470.8	469.7	466.7	461.7	451.6	419.6	361.6	289.0
550 K	h_M	718.4	718.3	718.2	717.9	717.1	715.8	717.6	705.4	693.5	674.4
	h_E	617.8	617.5	617.2	616.5	614.5	611.4	605.4	587.7	559.1	511.1
	h_P	604.4	604.1	603.5	602.6	600.0	595.9	587.8	563.1	521.1	458.8
600 K	h_M	877.2	877.2	877.0	876.8	876.2	875.2	873.2	867.3	858.3	844.1
	h_E	761.5	761.3	761.0	760.4	758.7	756.2	751.1	736.5	713.5	674.6
	h_P	745.8	745.4	744.9	744.1	741.9	738.5	731.9	712.1	679.9	628.8

Tabulated data based upon the equation of state of Younglove, B.A., and Ely, J.F., 1987, *J. Phys. Chem. Ref. Data*, **16(4)**: 577–798.

Psychrometric chart for water vapour in air (low temperature range)

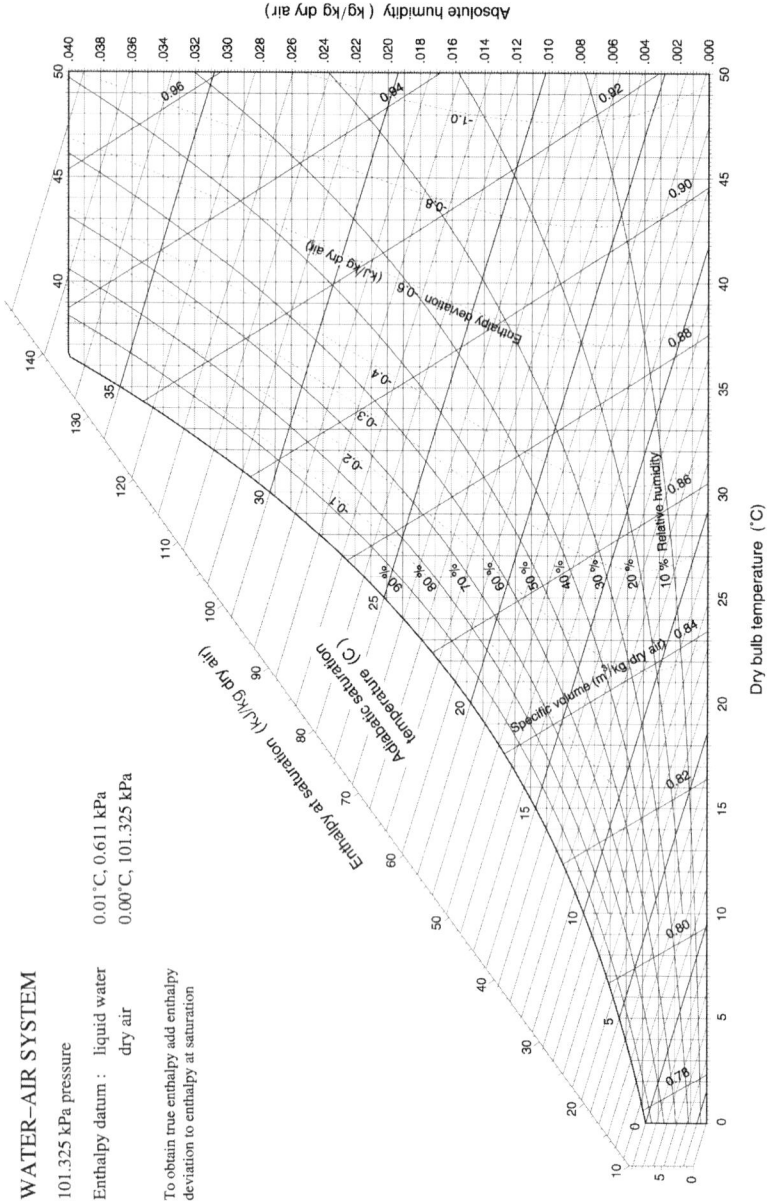

WATER–AIR SYSTEM

101.325 kPa pressure

Enthalpy datum : liquid water   0.01°C, 0.611 kPa
                  dry air   0.00°C, 101.325 kPa

To obtain true enthalpy add enthalpy
deviation to enthalpy at saturation

Chart reprinted from an extensive collection of psychrometric charts presented by Shallcross, D.C., 1997, *Handbook of Psychrometric Charts* (Chapman and Hall, London, UK).

Psychrometric chart for water vapour in air (high temperature range)

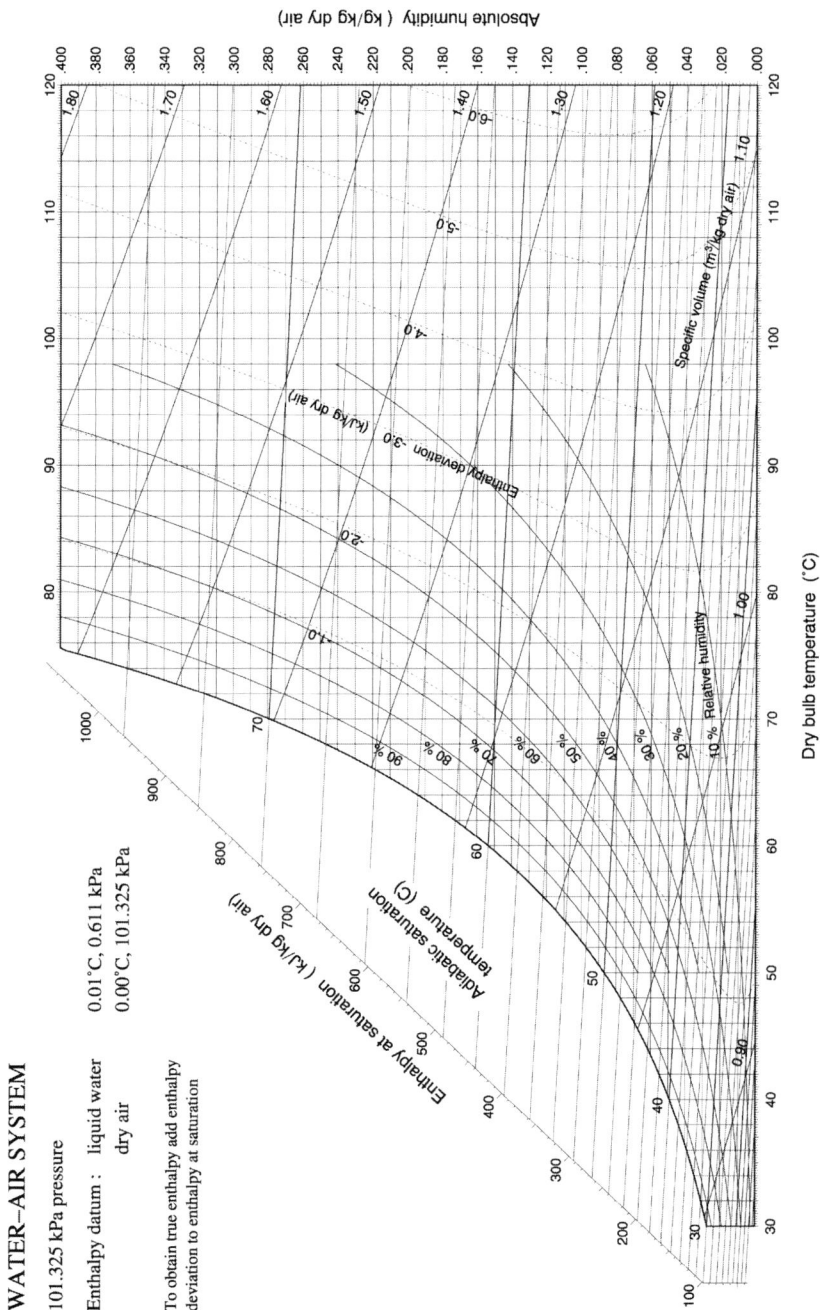

Chart reprinted from an extensive collection of psychrometric charts presented by Shallcross, D.C., 1997, *Handbook of Psychrometric Charts* (Chapman and Hall, London, UK).

Psychrometric chart for carbon tetrachloride vapour in air

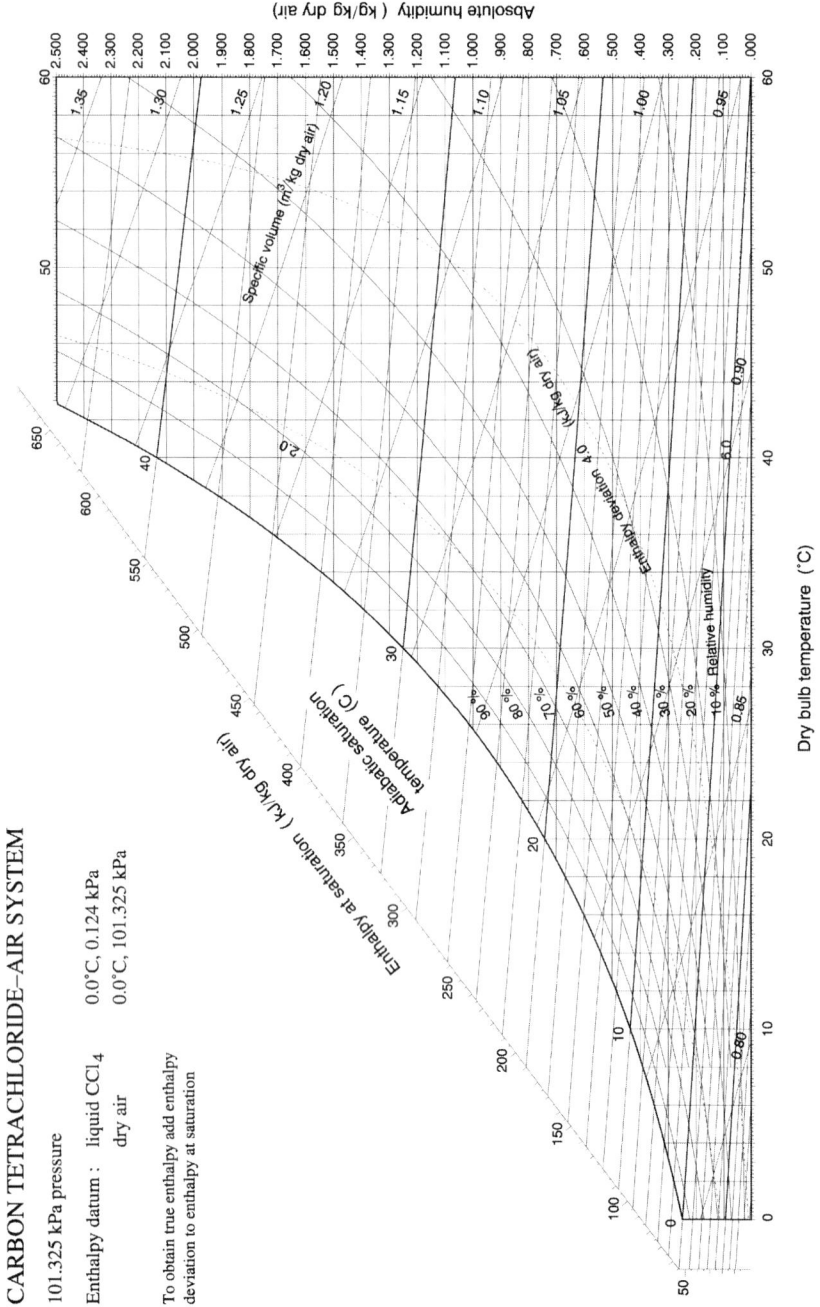

Chart reprinted from an extensive collection of psychrometric charts presented by Shallcross, D.C., 1997, *Handbook of Psychrometric Charts* (Chapman and Hall, London, UK).

Psychrometric chart for toluene vapour in nitrogen

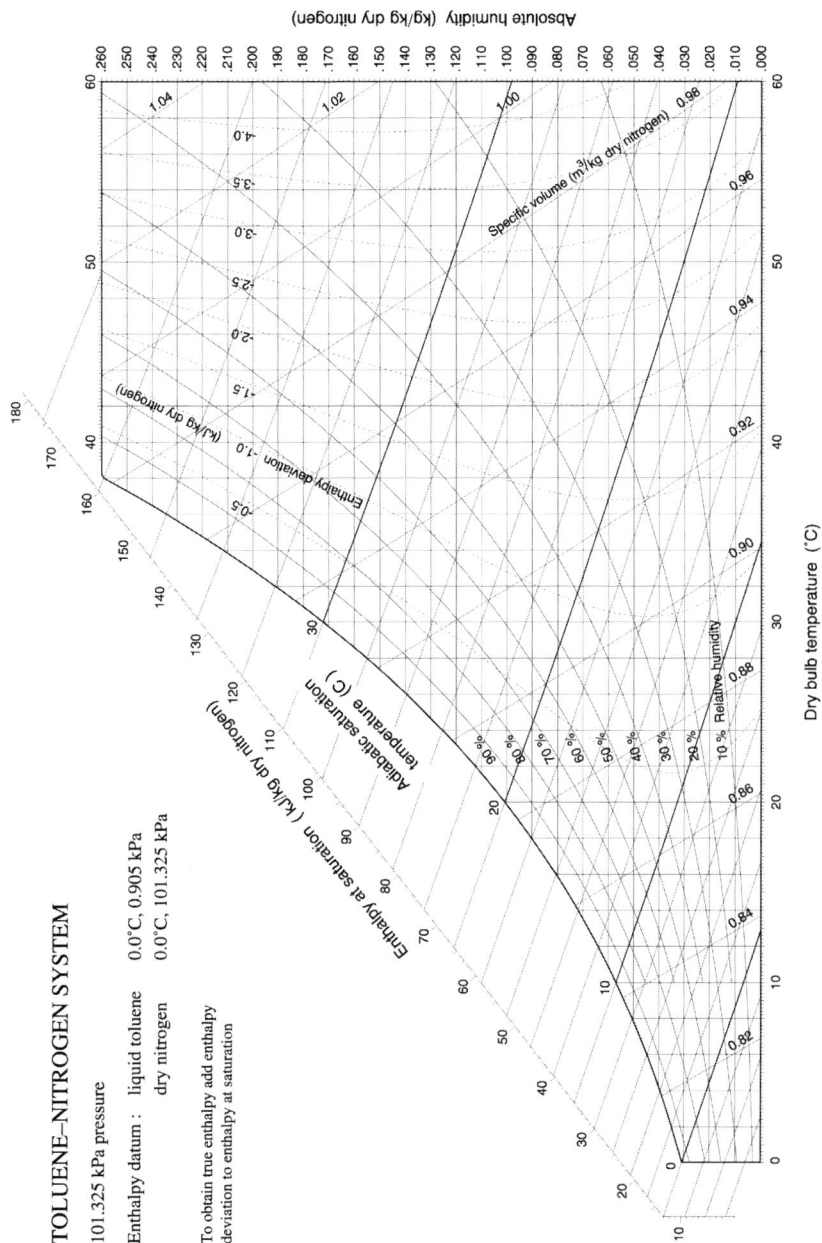

TOLUENE–NITROGEN SYSTEM

101.325 kPa pressure

Enthalpy datum : liquid toluene   0.0°C, 0.905 kPa
                 dry nitrogen   0.0°C, 101.325 kPa

To obtain true enthalpy add enthalpy
deviation to enthalpy at saturation

Chart reprinted from an extensive collection of psychrometric charts presented by Shallcross, D.C., 1997, *Handbook of Psychrometric Charts* (Chapman and Hall, London, UK).

41

Properties of dry air at 101.325 kPa

T ($^\circ$C)	z	ρ (kg m^{-3})	C_P (kJ kg^{-1} K^{-1})	C_V (kJ kg^{-1} K^{-1})	γ	h (kJ kg^{-1})	k (mW m^{-1} K^{-1})	μ (μPa s)	Pr
-100	0.99601	2.0197	1.008	0.716	1.408	100.7	16.02	11.75	0.739
-80	0.99723	1.8083	1.007	0.716	1.406	80.6	17.72	12.92	0.734
-60	0.99807	1.6373	1.006	0.716	1.405	60.4	19.39	14.05	0.729
-40	0.99865	1.4960	1.006	0.716	1.405	40.3	21.02	15.15	0.725
-20	0.99907	1.3772	1.006	0.717	1.403	20.1	22.61	16.21	0.721
0	0.99938	1.2759	1.006	0.717	1.403	0.0	24.16	17.23	0.717
20	0.99962	1.1886	1.007	0.718	1.403	20.1	25.68	18.22	0.715
40	0.99979	1.1125	1.008	0.719	1.402	40.3	27.16	19.19	0.712
60	0.99993	1.0456	1.008	0.720	1.400	60.4	28.61	20.13	0.709
80	1.00003	0.9863	1.009	0.721	1.399	80.6	30.03	21.04	0.707
100	1.00011	0.9333	1.011	0.724	1.397	100.8	31.42	21.93	0.706
120	1.00017	0.8858	1.013	0.726	1.395	121.1	32.79	22.80	0.705
140	1.00022	0.8429	1.016	0.729	1.394	141.4	34.12	23.64	0.704
160	1.00026	0.8039	1.019	0.732	1.393	161.7	35.44	24.47	0.704
180	1.00029	0.7684	1.023	0.735	1.392	182.1	36.73	25.29	0.704
200	1.00031	0.7359	1.026	0.738	1.390	202.7	38.01	26.08	0.704
220	1.00033	0.7061	1.029	0.742	1.387	223.2	39.26	26.86	0.704
240	1.00035	0.6785	1.033	0.746	1.385	243.8	40.50	27.63	0.704
260	1.00036	0.6531	1.036	0.750	1.383	264.5	41.72	28.39	0.705
280	1.00036	0.6295	1.041	0.754	1.381	285.2	42.93	29.13	0.706
300	1.00037	0.6075	1.046	0.758	1.379	306.1	44.13	29.86	0.707
320	1.00037	0.5870	1.050	0.762	1.378	327.0	45.31	30.57	0.709
340	1.00038	0.5679	1.055	0.767	1.375	348.1	46.48	31.28	0.710
360	1.00038	0.5499	1.059	0.772	1.372	369.3	47.64	31.98	0.711
380	1.00038	0.5331	1.064	0.777	1.369	390.5	48.79	32.67	0.712
400	1.00038	0.5172	1.069	0.782	1.367	411.8	49.93	33.35	0.714
420	1.00037	0.5023	1.073	0.786	1.365	433.2	51.06	34.02	0.715
440	1.00037	0.4882	1.078	0.791	1.362	454.7	52.18	34.68	0.716
460	1.00037	0.4749	1.082	0.796	1.360	476.4	53.30	35.33	0.717
480	1.00037	0.4623	1.088	0.801	1.358	498.0	54.41	35.98	0.719
500	1.00036	0.4503	1.093	0.806	1.356	519.9	55.51	36.62	0.721
520	1.00036	0.4390	1.097	0.810	1.354	541.8	56.60	37.25	0.722
540	1.00036	0.4282	1.102	0.815	1.352	563.8	57.69	37.87	0.723
560	1.00035	0.4179	1.106	0.819	1.349	585.9	58.77	38.49	0.724
580	1.00035	0.4081	1.111	0.824	1.348	608.0	59.85	39.10	0.726
600	1.00035	0.3988	1.115	0.828	1.347	630.3	60.92	39.71	0.727
620	1.00034	0.3898	1.119	0.832	1.345	652.7	61.99	40.31	0.728
640	1.00034	0.3813	1.124	0.837	1.343	675.1	63.05	40.90	0.729
660	1.00033	0.3731	1.127	0.841	1.340	697.6	64.11	41.49	0.730
680	1.00033	0.3653	1.133	0.846	1.339	720.2	65.16	42.07	0.731
700	1.00033	0.3578	1.136	0.849	1.338	742.9	66.21	42.65	0.732
720	1.00032	0.3506	1.140	0.853	1.337	765.7	67.26	43.23	0.732
740	1.00032	0.3437	1.143	0.857	1.335	788.6	68.30	43.80	0.733
760	1.00031	0.3370	1.147	0.861	1.332	811.5	69.34	44.36	0.734
780	1.00031	0.3306	1.151	0.865	1.332	834.4	70.38	44.92	0.735
800	1.00031	0.3245	1.154	0.868	1.330	857.5	71.41	45.47	0.735
850	1.00030	0.3100	1.162	0.876	1.327	915.4	73.99	46.85	0.736
900	1.00029	0.2968	1.170	0.884	1.324	973.8	76.54	48.19	0.737
950	1.00028	0.2847	1.178	0.891	1.321	1032.5	79.08	49.51	0.737
1000	1.00027	0.2735	1.184	0.898	1.319	1091.6	81.61	50.81	0.737

Enthalpy datum condition is dry air at 0.0°C and 101.325 kPa.

Tabulated data based upon the equation of state of Sychev, V.V., *et al.*, 1987, Thermodynamic Properties of Air (Hemisphere Publishing Corporation, New York, USA) and the correlations for thermal conductivity and viscosity of Kadoya, K., *et al.*, 1985, *J. Phys. Chem. Ref. Data*, **14(4)**: 947–970.

Enthalpy–concentration chart for sulphuric acid–water mixture

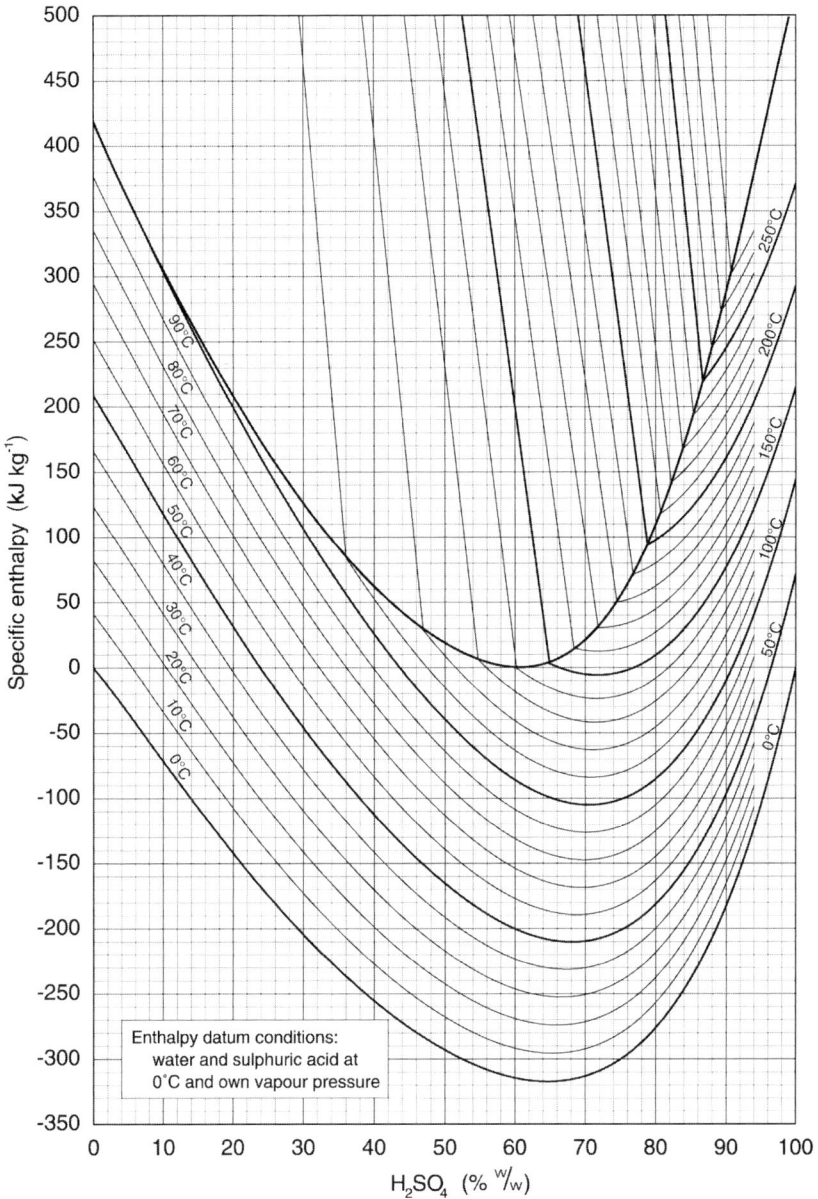

Enthalpy datum conditions:
water and sulphuric acid at
0°C and own vapour pressure

Specific enthalpy (kJ kg⁻¹) vs H_2SO_4 (% w/w)

Enthalpy–concentration chart for sodium hydroxide–water mixture

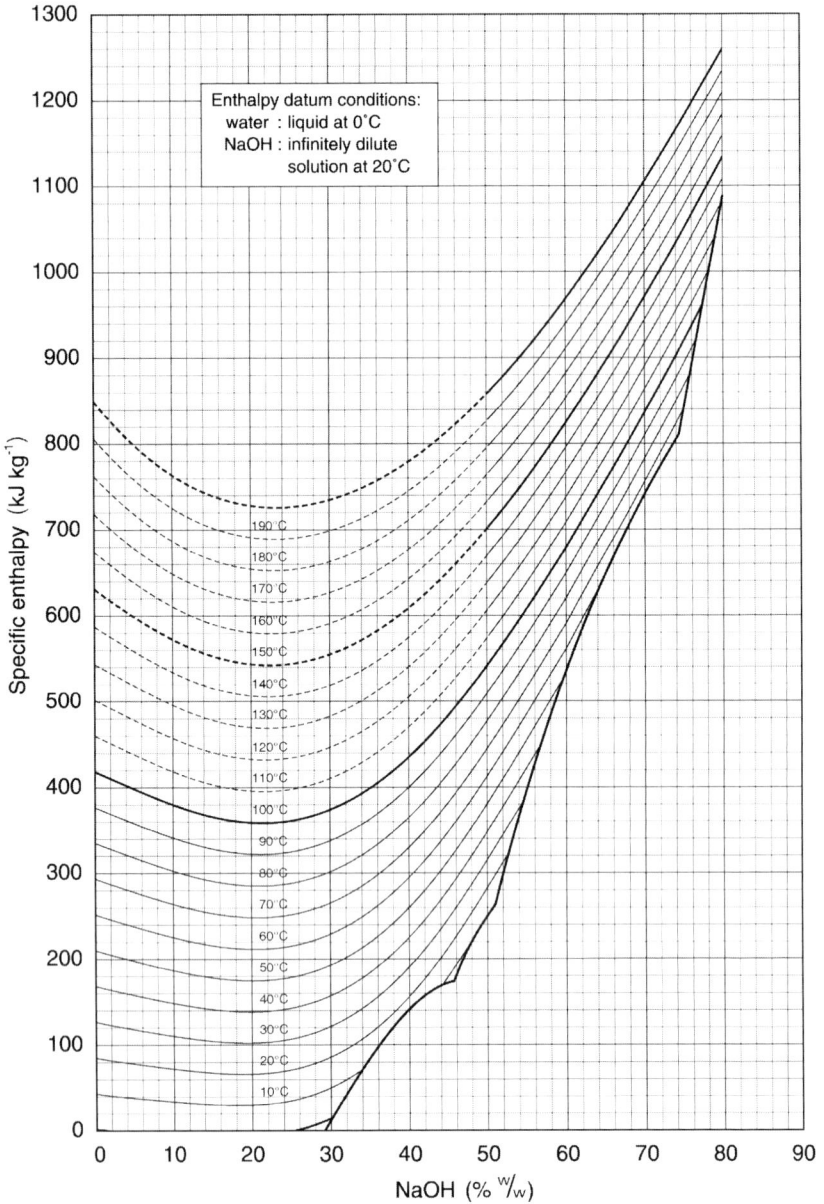

Enthalpy datum conditions:
water : liquid at 0°C
NaOH : infinitely dilute
solution at 20°C

Specific enthalpy (kJ kg⁻¹)

190°C
180°C
170°C
160°C
150°C
140°C
130°C
120°C
110°C
100°C
90°C
80°C
70°C
60°C
50°C
40°C
30°C
20°C
10°C

NaOH (% $^w/_w$)

Enthalpy – concentration chart for ethanol – water mixture

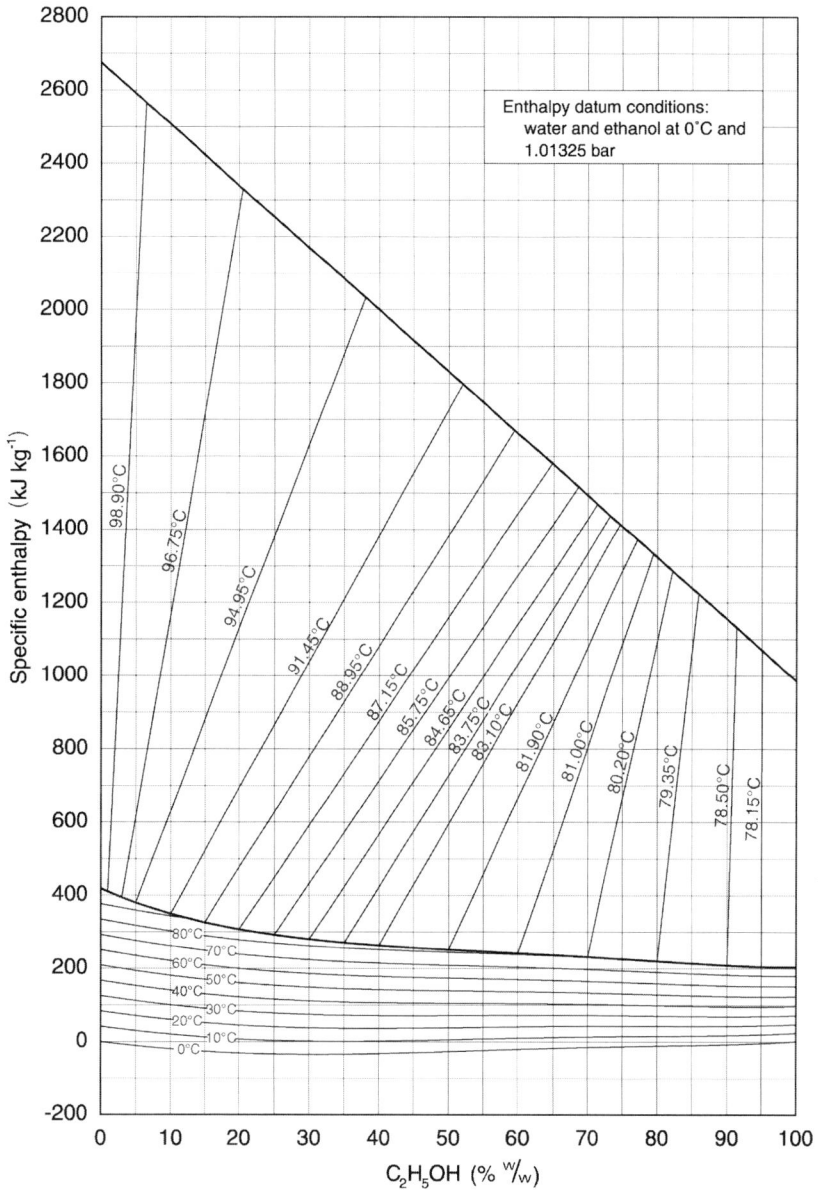

Enthalpy datum conditions:
water and ethanol at 0°C and
1.01325 bar

Specific enthalpy (kJ kg^{-1})

C_2H_5OH (% $^w/_w$)

98.90°C
96.75°C
94.95°C
91.45°C
88.95°C
87.15°C
85.75°C
84.65°C
83.75°C
83.10°C
81.90°C
81.00°C
80.20°C
79.35°C
78.50°C
78.15°C

80°C
70°C
60°C
50°C
40°C
30°C
20°C
10°C
0°C

Densities of selected gases at 101.325 kPa

T (K)	Ar	H$_2$	N$_2$	O$_2$	CO	CO$_2$	CH$_4$	C$_2$H$_6$	C$_3$H$_8$
200	2.441	0.1228	1.711	1.956	1.712	2.735	0.9838	1.882	
210	2.324	0.1169	1.629	1.862	1.629	2.596	0.9362	1.785	
220	2.218	0.1116	1.554	1.776	1.555	2.472	0.8929	1.699	
230	2.121	0.1068	1.486	1.699	1.487	2.360	0.8536	1.621	
240	2.032	0.1023	1.424	1.627	1.425	2.258	0.8176	1.550	2.317
250	1.950	0.09821	1.367	1.562	1.367	2.165	0.7845	1.486	2.215
260	1.875	0.09443	1.314	1.502	1.314	2.079	0.7541	1.426	2.121
270	1.805	0.09094	1.265	1.446	1.265	2.000	0.7259	1.372	2.036
280	1.740	0.08769	1.220	1.394	1.220	1.927	0.6998	1.321	1.958
290	1.680	0.08467	1.177	1.346	1.178	1.860	0.6755	1.274	1.887
300	1.624	0.08185	1.138	1.301	1.139	1.797	0.6528	1.231	1.820
310	1.571	0.07921	1.101	1.259	1.102	1.738	0.6316	1.190	1.759
320	1.522	0.07673	1.067	1.219	1.067	1.683	0.6118	1.152	1.701
330	1.476	0.07441	1.035	1.182	1.035	1.631	0.5932	1.117	1.648
340	1.432	0.07222	1.004	1.147	1.004	1.583	0.5756	1.083	1.598
350	1.391	0.07016	0.975	1.114	0.9756	1.537	0.5591	1.052	1.551
360	1.353	0.06821	0.948	1.083	0.9484	1.494	0.5435	1.022	1.506
370	1.316	0.06637	0.922	1.054	0.9226	1.453	0.5288	0.9942	1.464
380	1.281	0.06462	0.898	1.026	0.8983	1.415	0.5148	0.9678	1.425
390	1.248	0.06297	0.875	1.000	0.8753	1.378	0.5016	0.9427	1.388
400	1.217	0.06139	0.853	0.9752	0.8535	1.343	0.4890	0.9188	1.352
420	1.159	0.05847	0.813	0.9221	0.8126	1.279	0.4657	0.8747	1.286
440	1.107	0.05581	0.776	0.8833	0.7756	1.220	0.4444	0.8346	1.227
460	1.058	0.05339	0.742	0.8479	0.7419	1.167	0.4251	0.7981	1.173
480	1.014	0.05116	0.711	0.8126	0.7110	1.118	0.4073	0.7646	1.123
500	0.9736	0.04912	0.682	0.7800	0.6826	1.074	0.3910	0.7338	1.078
520	0.9360	0.04723	0.656	0.7500	0.6563	1.032	0.3760	0.7055	1.036
540	0.9013	0.04548	0.632	0.7222	0.6319	0.9937	0.3620	0.6792	0.9974
560	0.8693	0.04386	0.609	0.6964	0.6093	0.9581	0.3491	0.6549	0.9615
580	0.8393	0.04234	0.588	0.6724	0.5883	0.9250	0.3370	0.6322	0.9281
600	0.8114	0.04093	0.569	0.6499	0.5686	0.8941	0.3258	0.6111	0.8970
620	0.7850	0.03961	0.550	0.6290	0.5503	0.8652	0.3153		
640	0.7606	0.03838	0.533	0.6093	0.5331	0.8382			
660	0.7375	0.03721	0.517	0.5908	0.5169	0.8127			
680	0.7159	0.03612	0.502	0.5734	0.5017	0.7888			
700	0.6955	0.03509	0.487	0.5570	0.4874	0.7662			
720	0.6760	0.03411	0.474	0.5416	0.4738	0.7449			
740	0.6576	0.03319	0.461	0.5269	0.4610	0.7248			
760	0.6404	0.03232	0.449	0.5130	0.4489	0.7057			
780	0.6240	0.03149	0.438	0.4999	0.4375	0.6876			
800	0.6084	0.03070	0.427	0.4874	0.4266	0.6704			
850	0.5725	0.02890	0.401	0.4587	0.4014	0.6309			
900	0.5409	0.02729	0.379	0.4332	0.3793	0.5958			
950	0.5126	0.02586	0.359	0.4104	0.3591	0.5645			
1000	0.4866	0.02456	0.341	0.3899	0.3412	0.5362			

Density data expressed in kg m^{-3}.

Specific heat capacities of selected gases at 101.325 kPa

T (K)	Ar	H_2	N_2	O_2	CO	CO_2	CH_4	C_2H_6	C_3H_8
200	0.524	16.08	1.043	0.915	1.043	0.771	2.106	1.448	
210	0.523	15.92	1.042	0.915	1.043	0.775	2.110	1.474	
220	0.523	15.75	1.042	0.914	1.042	0.781	2.116	1.501	
230	0.523	15.61	1.042	0.914	1.042	0.788	2.123	1.530	
240	0.523	15.46	1.041	0.915	1.042	0.796	2.133	1.560	1.483
250	0.522	15.33	1.041	0.915	1.042	0.805	2.145	1.592	1.516
260	0.522	15.21	1.041	0.916	1.042	0.814	2.159	1.625	1.549
270	0.522	15.11	1.041	0.916	1.042	0.824	2.176	1.659	1.586
280	0.522	15.01	1.041	0.917	1.042	0.833	2.194	1.694	1.623
290	0.522	14.93	1.041	0.919	1.042	0.843	2.214	1.731	1.664
300	0.521	14.85	1.041	0.920	1.042	0.853	2.236	1.768	1.704
310	0.521	14.81	1.041	0.921	1.042	0.863	2.260	1.806	1.747
320	0.521	14.76	1.041	0.923	1.042	0.872	2.285	1.845	1.789
330	0.521	14.72	1.041	0.925	1.043	0.881	2.312	1.885	1.832
340	0.521	14.68	1.041	0.927	1.044	0.890	2.341	1.925	1.876
350	0.521	14.63	1.041	0.929	1.044	0.899	2.370	1.966	1.919
360	0.521	14.62	1.042	0.931	1.045	0.908	2.401	2.008	1.963
370	0.521	14.60	1.042	0.934	1.045	0.916	2.433	2.050	2.007
380	0.521	14.59	1.043	0.936	1.046	0.925	2.466	2.092	2.051
390	0.521	14.57	1.043	0.939	1.047	0.933	2.500	2.135	2.094
400	0.521	14.55	1.044	0.942	1.048	0.941	2.534	2.178	2.138
420	0.521	14.54	1.046	0.948	1.051	0.957	2.605	2.264	2.224
440	0.521	14.53	1.047	0.953	1.054	0.973	2.677	2.350	2.308
460	0.521	14.53	1.050	0.959	1.057	0.987	2.751	2.435	2.392
480	0.521	14.53	1.052	0.966	1.060	1.002	2.826	2.517	2.474
500	0.521	14.53	1.055	0.972	1.064	1.015	2.901	2.597	2.553
520	0.521	14.53	1.058	0.978	1.069	1.029	2.976	2.675	2.630
540	0.521	14.53	1.062	0.985	1.073	1.041	3.051	2.752	2.705
560	0.521	14.53	1.066	0.991	1.077	1.054	3.125	2.826	2.780
580	0.521	14.54	1.069	0.997	1.082	1.065	3.200	2.899	2.850
600	0.520	14.55	1.074	1.003	1.087	1.077	3.273	2.968	2.921
620	0.520	14.56	1.078	1.009	1.082	1.088	3.346		
640	0.520	14.57	1.082	1.015	1.077	1.098			
660	0.520	14.59	1.087	1.021	1.171	1.108			
680	0.520	14.60	1.092	1.026	1.142	1.118			
700	0.520	14.61	1.096	1.031	1.113	1.127			
720	0.520	14.62	1.101	1.036	1.098	1.136			
740	0.520	14.64	1.106	1.041	1.083	1.145			
760	0.520	14.66	1.111	1.045	1.176	1.154			
780	0.520	14.68	1.116	1.050	1.158	1.162			
800	0.520	14.70	1.120	1.054	1.139	1.169			
850	0.520	14.76	1.105	1.065	1.151	1.187			
900	0.520	14.83	1.116	1.074	1.163	1.204			
950	0.520	14.90	1.127	1.082	1.174	1.219			
1000	0.520	14.99	1.137	1.090	1.185	1.234			

Specific heat capacity data expressed in kJ kg^{-1} K^{-1}.

Thermal conductivities capacities of selected gases at 101.325 kPa

T (K)	Ar	H_2	N_2	O_2	CO	CO_2	CH_4	C_2H_6	C_3H_8
200	12.46	131	18.72	18.48	17.38		21.56		
210	13.03	137	19.51	19.29	18.21		22.75		
220	13.59	142	20.29	20.10	19.02		23.97		
230	14.14	147	21.05	20.89	19.82		25.22		
240	14.68	152	21.79	21.67	20.61	11.96	26.51	14.4	
250	15.22	157	22.52	22.44	21.38	12.75	27.83	15.5	12.6
260	15.75	162	23.24	23.20	22.15	13.55	29.18	16.6	13.7
270	16.27	167	23.95	23.95	22.90	14.35	30.57	17.8	14.7
280	16.78	172	24.65	24.69	23.64	15.16	31.99	19.0	15.8
290	17.29	178	25.33	25.42	24.38	15.98	33.43	20.2	17.0
300	17.79	183	26.01	26.15	25.11	16.79	34.91	21.5	18.1
310	18.27	187	26.68	26.88	25.83	17.61	36.42	22.8	19.3
320	18.74	191	27.33	27.60	26.54	18.43	37.96	24.1	20.5
330	19.22	196	27.98	28.32	27.25	19.26	39.52	25.4	21.7
340	19.69	200	28.63	29.03	27.95	20.09	41.11	26.8	23.0
350	20.15	204	29.26	29.74	28.64	20.92	42.73	28.2	24.2
360	20.61	208	29.89	30.44	29.33	21.75	44.37	29.7	25.5
370	21.06	213	30.51	31.15	30.02	22.58	46.04	31.2	26.9
380	21.51	217	31.13	31.85	30.70	23.41	47.73	32.6	28.2
390	21.96	222	31.74	32.55	31.38	24.24	49.45	34.2	29.6
400	22.40	226	32.35	33.24	32.06	25.07	51.19	35.7	31.0
420	23.27	234	33.56	34.63	33.41	26.73	54.72	38.9	33.8
440	24.13	243	34.74	36.01	34.75	28.39	58.34	42.1	36.7
460	24.97	251	35.92	37.38	36.08	30.04	62.03	45.5	39.7
480	25.79	258	37.09	38.74	37.42	31.69	65.79	48.9	42.8
500	26.60	266	38.25	40.10	38.75	33.32	69.59	52.4	45.9
520	27.40	274	39.40	41.45	40.08	34.95	73.42	56.0	49.1
540	28.18	281	40.55	42.79	41.41	36.57	77.34	59.6	52.4
560	28.95	289	41.70	44.12	42.74	38.18	81.31	63.4	55.7
580	29.71	297	42.83	45.44	44.08	39.78	85.33	67.1	59.2
600	30.45	305	43.97	46.74	45.41	41.36	89.39		62.6
620	31.18	312	45.10	48.04	46.74	42.94	93.50		66.2
640	31.90	319	46.23	49.32	48.08	44.50	97.64		69.8
660	32.61	327	47.36	50.58	49.43	46.04	101.8		73.5
680	33.31	334	48.48	51.84	50.76	47.57	106.0		77.2
700	34.00	342	49.60	53.08	52.09	49.10	110.3		81.0
720	34.67	349	50.71	54.32	53.41	50.63	114.5		85.0
740	35.34	356	51.82	55.54	54.73	52.16	118.8		88.9
760	36.00	364	52.93	56.75	56.04	53.68	123.1		93.0
780	36.66	371	54.03	57.95	57.35	55.19	127.4		97.2
800	37.30	378	55.13	59.14	58.65	56.69	131.8		101.4
850	38.88	395	57.86	62.07	61.88	60.40	142.7		
900	40.43	412	60.55	64.93	65.07	64.05	153.6		
950	41.97	430	63.21	67.74	68.23	67.62	164.5		
1000	43.47	448	65.84	70.48	71.33	71.12	175.4		

Thermal conductivity data expressed in mW m^{-1} K^{-1}.

Viscosities of selected gases at 101.325 kPa

T (K)	Ar	H_2	N_2	O_2	CO	CO_2	CH_4	C_2H_6	C_3H_8
200	15.99	6.83	12.89	14.80	12.86	10.01	7.78	6.31	
210	16.70	7.05	13.43	15.41	13.36	10.53	8.13	6.63	
220	17.41	7.27	13.97	16.03	13.85	11.04	8.48	6.95	
230	18.12	7.48	14.49	16.63	14.34	11.56	8.82	7.27	
240	18.82	7.69	15.00	17.23	14.82	12.07	9.16	7.58	
250	19.51	7.90	15.51	17.82	15.30	12.57	9.49	7.89	
260	20.18	8.11	16.00	18.40	15.77	13.07	9.82	8.20	
270	20.83	8.32	16.49	18.98	16.23	13.57	10.15	8.50	
280	21.46	8.52	16.96	19.55	16.69	14.06	10.48	8.80	7.70
290	22.09	8.73	17.44	20.11	17.14	14.55	10.80	9.10	7.95
300	22.71	8.93	17.90	20.66	17.59	15.04	11.12	9.39	8.21
310	23.33	9.13	18.36	21.21	18.03	15.52	11.43	9.68	8.46
320	23.94	9.32	18.80	21.76	18.47	16.00	11.74	9.97	8.71
330	24.58	9.52	19.25	22.29	18.90	16.48	12.05	10.26	8.96
340	25.21	9.71	19.69	22.83	19.33	16.95	12.35	10.54	9.21
350	25.83	9.91	20.12	23.35	19.75	17.41	12.65	10.82	9.46
360	26.44	10.10	20.54	23.87	20.17	17.88	12.94	11.10	9.71
370	27.04	10.29	20.96	24.38	20.58	18.34	13.24	11.38	9.95
380	27.64	10.47	21.38	24.89	20.99	18.79	13.53	11.65	10.20
390	28.22	10.66	21.79	25.39	21.39	19.25	13.81	11.92	10.44
400	28.80	10.85	22.19	25.89	21.78	19.70	14.08	12.19	10.69
420	29.93	11.21	22.99	26.86	22.56	20.58	14.64	12.72	11.17
440	31.03	11.57	23.77	27.82	23.32	21.46	15.18	13.25	11.65
460	32.11	11.92	24.54	28.75	24.07	22.32	15.71	13.76	12.13
480	33.16	12.27	25.29	29.67	24.79	23.17	16.21	14.27	12.61
500	34.18	12.61	26.02	30.56	25.50	24.00	16.70	14.78	13.08
520	35.17	12.95	26.75	31.44	26.19	24.82	17.18	15.28	13.56
540	36.15	13.28	27.46	32.30	26.87	25.63	17.66	15.77	14.03
560	37.10	13.61	28.16	33.14	27.53	26.43	18.17	16.26	14.49
580	38.04	13.93	28.90	33.97	28.18	27.21		16.74	14.96
600	38.95	14.25	29.58	34.78	28.81	27.98		17.23	15.43
620	39.85	14.56	30.24	35.58	29.43	28.74		17.71	15.89
640	40.73	14.87	30.89	36.37	30.03	29.49		18.18	16.36
660	41.60	15.18	31.54	37.14	30.62	30.23		18.66	16.82
680	42.46	15.48	32.17	37.90	31.21	30.96		19.14	17.28
700	43.30	15.78	32.80	38.65	31.77	31.67		19.61	17.74
720	44.13	16.08	33.42	39.38	32.33	32.38		20.09	18.21
740	44.96	16.37	34.03	40.11	32.88	33.07		20.57	18.67
760	45.78	16.66	34.63	40.83	33.42	33.76		21.04	19.13
780	46.59	16.95	35.23	41.54	33.94	34.43		21.53	19.60
800	47.44	17.24	35.82	42.27	34.46	35.10		22.01	20.06
850	49.43	17.95	37.28	44.03	35.72	36.72			
900	51.38	18.65	38.72	45.75	36.94	38.29			
950	53.26	19.34	40.15	47.43	38.12	39.81			
1000	55.14	20.02	41.58	49.08	39.27	41.28			

Viscosity data expressed in μPa s.

Antoine vapour pressure equation coefficient data

The Antoine equation may be used to estimate the vapour pressure of a substance as a function of its temperature. The equation is:

$$\log_{10} p = a - \frac{b}{c + T}$$

where, p is the vapour pressure and T is the temperature. If p is expressed in pascals and T in kelvins then for the compounds listed the coefficients a, b and c have the values shown below. For most compounds the estimated vapour pressure will be within 2% of the actual values.

When applying the equation it must be remembered that the values for a, b and c for a given compound are only valid for the temperature range between T_{min} and T_{max}. On no account should the equation be extrapolated beyond these limits.

Compound		a	b	c	T_{min} (K)	T_{max} (K)
alkanes						
methane	CH_4	8.997	447.4	0.36	95	185
ethane	C_2H_6	9.025	685.4	−14.09	95	295
propane	C_3H_8	9.032	842.8	−21.25	250	365
n-butane	C_4H_{10}	9.089	1005	−26.68	150	400
iso-butane	C_4H_{10}	8.990	928.6	−28.11	200	395
n-pentane	C_5H_{12}	9.058	1099	−37.95	245	465
n-hexane	C_6H_{14}	9.012	1172	−49.26	180	480
n-heptane	C_7H_{16}	9.130	1322	−51.25	195	530
n-octane	C_8H_{18}	9.182	1434	−55.42	230	565
n-nonane	C_9H_{20}	9.378	1644	−48.72	300	580
n-decane	$C_{10}H_{22}$	9.242	1611	−67.20	250	615
cyclobutane	C_4H_8	9.181	1088	−25.31	185	440
cyclopentane	C_5H_{10}	9.124	1183	−35.25	185	500
cyclohexane	C_6H_{12}	9.131	1307	−37.36	285	540
cycloheptane	C_7H_{14}	9.078	1391	−50.55	270	580
alkenes						
ethene	C_2H_4	9.029	633.1	−11.96	110	270
propene	C_3H_6	9.100	845.9	−18.99	145	360
1-butene	C_4H_8	9.000	928.7	−34.63	180	415
1-pentene	C_5H_{10}	9.041	1074	−36.96	200	440
propadiene	C_3H_4	9.774	1321	39.09	245	380
alkynes						
ethyne	C_2H_2	9.501	830.0	−3.25	200	305
propyne	C_3H_4	9.200	912.7	−32.58	175	385
1-butyne	C_4H_6	9.263	1058	−32.84	190	440
2-butyne	C_4H_6	8.979	992.1	−50.20	250	485
aromatic compounds						
benzene	C_6H_6	9.176	1301	−41.73	280	550
toluene	$C_6H_5CH_3$	9.257	1456	−41.93	260	600
o-xylene	$C_6H_4(CH_3)_2$	9.218	1536	−53.11	260	610
m-xylene	$C_6H_4(CH_3)_2$	9.262	1548	−48.83	240	600
p-xylene	$C_6H_4(CH_3)_2$	9.123	1467	−55.55	290	575
ethylbenzene	$C_6H_5C_2H_5$	9.216	1513	−50.25	280	610
naphthalene	$C_{10}H_8$	9.226	1806	−63.19	355	725

Antoine vapour pressure equation coefficient data (cont.)

Compound		a	b	c	T_{min} (K)	T_{max} (K)
alcohols						
methanol	CH_3OH	10.197	1572	-35.07	245	510
ethanol	C_2H_5OH	10.142	1556	-48.73	260	500
1-propanol	C_3H_7OH	9.959	1516	-65.22	260	480
2-propanol	C_3H_7OH	9.894	1413	-66.75	265	485
isobutanol	C_4H_9OH	9.411	1239	-100.00	325	505
n-butanol	C_4H_9OH	9.852	1540	-73.89	325	480
1-pentanol	$C_5H_{11}OH$	9.391	1367	-99.49	300	585
1-hexanol	$C_6H_{13}OH$	9.707	1641	-82.41	280	485
1-heptanol	$C_7H_{15}OH$	9.318	1446	-114.19	365	465
1-octanol	$C_8H_{17}OH$	9.350	1542	-113.49	310	485
cyclohexanol	$C_6H_{11}OH$	9.488	1512	-96.73	310	535
phenol	C_6H_5OH	9.706	1815	-68.63	325	660
other organic compounds						
formaldehyde	$HCHO$	9.556	1074	-18.25	190	390
acetaldehyde	CH_3CHO	9.400	1222	-16.52	255	410
benzaldehyde	C_6H_5CHO	9.367	1714	-59.21	290	690
acetone	CH_3COCH_3	9.409	1301	-34.04	185	505
methyl ethyl ketone	$C_2H_5COCH_3$	9.344	1360	-38.75	190	515
diethyl ketone	$C_2H_5COC_2H_5$	9.347	1441	-43.35	235	540
dimethyl ether	CH_3OCH_3	9.309	983.0	-20.48	140	400
methyl ethyl ether	$CH_3OC_2H_5$	8.733	757.4	-77.22	255	315
diethyl ether	$C_2H_5OC_2H_5$	9.073	1070	-44.75	185	440
formic acid	$HCOOH$	10.239	2047	18.00	300	375
acetic acid	CH_3COOH	9.874	1770	-27.78	300	580
hydrogen cyanide	HCN	10.198	1670	22.59	270	445
nitromethane	CH_3NO_2	9.579	1549	-35.79	255	430
methyl mercaptan	CH_3SH	9.075	953.0	-45.63	195	445
chloroform	$CHCl_3$	9.101	1170	-48.54	210	480
carbon tetrachloride	CCl_4	9.099	1266	-40.83	250	540
inorganic compounds						
carbon dioxide	CO	9.666	840.4	-3.86	220	300
hydrogen bromide	HBr	9.161	756.9	-24.76	185	360
hydrogen chloride	HCl	9.404	790.2	-8.729	165	310
hydrogen fluoride	HF	10.609	2035	70.05	260	430
hydrogen iodide	HI	9.209	941.9	-13.68	225	420
nitric acid	HNO_3	9.491	1336	-58.25	235	375
nitrogen dioxide	NO_2	12.149	2528	59.45	265	430
nitrous oxide	N_2O	9.148	644.2	-29.05	185	305
sulphur dioxide	SO_2	9.474	1017	-35.72	210	430
hydrogen sulphide	H_2S	9.254	808.9	-22.49	190	370

Data regressed by the author from various sources.

Unit conversions

Area

$1\ cm^2\ :\ 1.000\ 0 \times 10^{-4}\ m^2$
$1\ in^2\ :\ 6.451\ 6 \times 10^{-4}\ m^2$
$1\ ft^2\ :\ 9.290\ 3 \times 10^{-2}\ m^2$
$1\ yd^2\ :\ 8.361\ 3 \times 10^{-1}\ m^2$
$1\ acre\ :\ 4.046\ 9 \times 10^{3}\ m^2$
$1\ mile^2\ :\ 2.590\ 0 \times 10^{6}\ m^2$

Density

$1\ g/cm^3\ :\ 1.000\ 0 \times 10^{3}\ kg/m^3$
$1\ lb/ft^3\ :\ 1.601\ 8 \times 10\ kg/m^3$
$1\ lb/U.K.gal\ :\ 9.977\ 6 \times 10\ kg/m^3$
$1\ lb/U.S.gal\ :\ 1.198\ 3 \times 10^{2}\ kg/m^3$

Energy

$1\ cal\ :\ 4.186\ 8\ J$
$1\ kcal\ :\ 4.186\ 8 \times 10^{3}\ J$
$1\ Btu\ :\ 1.055\ 1 \times 10^{3}\ J$
$1\ erg\ :\ 1.000\ 0 \times 10^{-7}\ J$
$1\ hp\ h\ (metric)\ :\ 2.647\ 8 \times 10^{6}\ J$
$1\ hp\ h\ (British)\ :\ 2.684\ 5 \times 10^{6}\ J$
$1\ ft\ lb_f\ :\ 1.355\ 8\ J$

Force

$1\ dyne\ :\ 1.000 \times 10^{-5}\ N$
$1\ lb_f\ :\ 4.448\ 2\ N$

Heat flux

$1\ cal\ s^{-1}\ cm^{-2}\ :\ 4.186\ 8 \times 10^{4}\ W\ m^{-2}$
$1\ kcal\ h^{-1}\ m^{-2}\ :\ 1.163\ 0\ W\ m^{-2}$
$1\ Btu\ h^{-1}\ ft^{-2}\ :\ 3.154\ 6\ W\ m^{-2}$
$1\ kcal\ h^{-1}\ ft^{-2}\ :\ 1.251\ 8 \times 10\ W\ m^{-2}$

Heat transfer coefficient

$1\ cal\ s^{-1}\ cm^{-2}\ °C^{-1}\ :\ 4.186\ 8 \times 10^{4}\ W\ m^{-2}\ K^{-1}$
$1\ kcal\ h^{-1}\ m^{-2}\ °C^{-1}\ :\ 1.163\ 0\ W\ m^{-2}\ K^{-1}$
$1\ Btu\ h^{-1}\ ft^{-2}\ °F^{-1}\ :\ 5.678\ 3\ W\ m^{-2}\ K^{-1}$
$1\ kcal\ h^{-1}\ ft^{-2}\ °C^{-1}\ :\ 1.251\ 8 \times 10\ W\ m^{-2}\ K^{-1}$

Length

$1\ Ångstrom\ :\ 1.000\ 0 \times 10^{-10}\ m$
$1\ micron\ :\ 1.000\ 0 \times 10^{-6}\ m$
$1\ in\ :\ 2.540\ 0 \times 10^{-2}\ m$
$1\ ft\ :\ 3.048\ 0 \times 10^{-1}\ m$
$1\ yd\ :\ 9.144\ 0 \times 10^{-1}\ m$
$1\ fathom\ :\ 1.828\ 8\ m$
$1\ mile\ :\ 1.609\ 3 \times 10^{3}\ m$
$1\ light\ year\ :\ 9.460\ 5 \times 10^{15}\ m$

Mass

$1\ grain\ :\ 6.479\ 9 \times 10^{-5}\ kg$
$1\ oz\ :\ 2.835\ 0 \times 10^{-2}\ kg$
$1\ lb\ :\ 4.535\ 9 \times 10^{-1}\ kg$
$1\ cwt\ (long)\ :\ 5.080\ 2 \times 10\ kg$
$1\ tonne\ :\ 1.000\ 0 \times 10^{3}\ kg$
$1\ ton\ (long)\ :\ 1.016\ 0 \times 10^{3}\ kg$

Power

$1\ cal\ s^{-1}\ :\ 4.186\ 8\ W$
$1\ kcal\ h^{-1}\ :\ 1.163\ 0\ W$
$1\ Btu\ s^{-1}\ :\ 1.055\ 1 \times 10^{3}\ W$
$1\ Btu\ h^{-1}\ :\ 2.930\ 7 \times 10^{-1}\ W$
$1\ erg\ s^{-1}\ :\ 1.000\ 0 \times 10^{-7}\ W$
$1\ hp\ :\ 7.355\ 0 \times 10^{2}\ W$
(metric)
$1\ hp\ :\ 7.457\ 0 \times 10^{2}\ W$
(British)

Pressure

$1\ std\ atmosphere\ :\ 1.013\ 25 \times 10^{5}\ Pa$
$1\ bar\ :\ 1.000\ 0 \times 10^{5}\ Pa$
$1\ lb_f\ ft^{-2}\ :\ 4.788\ 0 \times 10\ Pa$
$1\ lb_f in^{-2}\ :\ 6.894\ 8 \times 10^{3}\ Pa$
(1 psi)
$1\ in\ water\ :\ 2.490\ 9 \times 10^{2}\ Pa$
$1\ ft\ water\ :\ 2.989\ 0 \times 10^{3}\ Pa$
$1\ mm\ Hg\ :\ 1.333\ 2 \times 10^{2}\ Pa$
$1\ in\ Hg\ :\ 3.386\ 4 \times 10^{3}\ Pa$

Specific enthalpy

$1\ cal/g\ :\ 4.186\ 8 \times 10^{3}\ J\ kg^{-1}$
$1\ Btu/lb\ :\ 2.326\ 0 \times 10^{3}\ J\ kg^{-1}$

Specific heat capacity

$1\ cal\ g^{-1}\ °C^{-1}\ :\ 4.186\ 8 \times 10^{3}\ J\ kg^{-1}\ K^{-1}$
$1\ Btu\ lb^{-1}\ °F^{-1}\ :\ 4.186\ 8 \times 10^{3}\ J\ kg^{-1}\ K^{-1}$

Specific volume

$1\ ft^3\ lb^{-1}\ :\ 6.242\ 8 \times 10^{-2}\ m^3\ kg^{-1}$
$1\ ft^3\ kg^{-1}\ :\ 2.831\ 7 \times 10^{-2}\ m^3\ kg^{-1}$

Thermal conductivity

$1\ cal\ s^{-1}\ cm^{-2}\ :\ 4.186\ 8 \times 10^{2}\ W\ m^{-1}\ K^{-1}$
$(°C\ cm^{-1})^{-1}$
$1\ kcal\ h^{-1}\ m^{-2}\ :\ 1.163\ 0\ W\ m^{-1}\ K^{-1}$
$(°C\ m^{-1})^{-1}$
$1\ Btu\ h^{-1}\ ft^{-2}\ :\ 1.730\ 7\ W\ m^{-1}\ K^{-1}$
$(°F\ ft^{-1})^{-1}$
$1\ Btu\ h^{-1}\ ft^{-2}\ :\ 1.442\ 3 \times 10^{-1}\ W\ m^{-1}\ K^{-1}$
$(°F\ in^{-1})^{-1}$
$1\ kcal\ h^{-1}\ ft^{-2}\ :\ 3.815\ 6\ W\ m^{-1}\ K^{-1}$
$(°C\ ft^{-1})^{-1}$

Time

$1\ min\ :\ 6.000\ 0 \times 10\ s$
$1\ h\ :\ 3.600\ 0 \times 10^{3}\ s$
$1\ day\ :\ 8.640\ 0 \times 10^{4}\ s$

Velocity

$1\ ft\ s^{-1}\ :\ 3.048\ 0 \times 10^{-1}\ m\ s^{-1}$
$1\ ft\ h^{-1}\ :\ 8.466\ 7 \times 10^{-5}\ m\ s^{-1}$
$1\ mile\ h^{-1}\ :\ 4.470\ 4 \times 10^{-1}\ m\ s^{-1}$
$1\ knot\ :\ 5.144\ 4 \times 10^{-1}\ m\ s^{-1}$

Viscosity, absolute

1 g cm^{-1} s^{-1} : 1.000 0 × 10^{-1} kg m^{-1} s^{-1}
 (1 poise)
1 kg m^{-1} h^{-1} : 2.777 8 × 10^{-4} kg m^{-1} s^{-1}
1 lb ft^{-1} s^{-1} : 1.488 2 kg m^{-1} s^{-1}
1 lb ft^{-1} h^{-1} : 4.133 8 × 10^{-4} kg m^{-1} s^{-1}

Viscosity, kinematic

1 cm^2 s^{-1} : 1.000 0 × 10^{-4} m^2 s^{-1}
 (1 stokes)
1 m^2 h^{-1} : 2.777 8 × 10^{-4} m^2 s^{-1}
1 ft^2 s^{-1} : 9.290 3 × 10^{-2} m^2 s^{-1}
1 ft^2 h^{-1} : 2.580 6 × 10^{-5} m^2 s^{-1}

Volume

1 cm^3 : 1.000 0 × 10^{-6} m^3
1 litre : 1.000 0 × 10^{-3} m^3
1 in^3 : 1.638 7 × 10^{-5} m^3
1 ft^3 : 2.831 7 × 10^{-2} m^3
1 yd^3 : 7.645 5 × 10^{-1} m^3
1 U.K.gal : 4.546 1 × 10^{-3} m^3
1 U.S.gal : 3.785 4 × 10^{-3} m^3
1 bbl : 1.589 9 × 10^{-1} m^3
1 acre-ft : 1.233 5 × 10^3 m^3

Volumetric flowrate

1 ft^3 s^{-1} : 2.831 7 × 10^{-2} m^3 s^{-1}
1 ft^3 min^{-1} : 4.719 5 × 10^{-4} m^3 s^{-1}
1 U.K.gal min^{-1} : 7.576 8 × 10^{-5} m^3 s^{-1}
1 U.S.gal min^{-1} : 6.309 0 × 10^{-5} m^3 s^{-1}

Temperature conversions

K	°C	°R	°F
0.00	− 273.15	0.00	− 459.67
100.00	− 173.15	180.00	− 279.67
200.00	− 73.15	360.00	− 99.67
255.37	− 17.78	459.67	0.00
273.15	0.00	491.67	32.00
298.15	25.00	536.67	77.00
300.00	26.85	540.00	80.33
366.48	93.33	659.67	200.00
373.15	100.00	671.67	212.00
400.00	126.85	720.00	260.33
473.15	200.00	851.67	392.00
477.59	204.44	859.67	400.00
500.00	226.85	900.00	440.33
533.15	260.00	959.67	500.00
573.15	300.00	1031.67	572.00
600.00	326.85	1080.00	620.33
673.15	400.00	1211.67	752.00
700.00	426.85	1260.00	800.33
773.15	500.00	1391.67	932.00
800.00	526.85	1440.00	980.33
810.93	537.78	1459.67	1000.00
900.00	626.85	1620.00	1160.33
1000.00	726.85	1800.00	1340.33
1100.00	826.85	1980.00	1520.33
1200.00	926.85	2160.00	1700.33
1273.15	1000.00	2291.67	1832.00
1300.00	1026.85	2340.00	1880.33
1366.48	1093.33	2459.67	2000.00
1400.00	1126.85	2520.00	2060.33
1500.00	1226.85	2700.00	2240.33

A more extensive collection of unit conversions is presented in the American Society for Testing and Materials Standard IEEE/ASTM SI 10-1997 Standard Practice for Use of the International System of Units (SI): The Modernized Metric System.

Critical point and standard heat of formation data

Compound		MW	T_C (K)	P_C (MPa)	$\hat{V}_C$ (m³ kmol⁻¹)	Z_C	ω	$\Delta \hat{h}_f^0$ (kJ mol⁻¹)	State
alkanes									
methane	CH_4	16.04	190.6	4.604	0.010	0.288	0.011	−74.9	g
ethane	C_2H_6	30.07	305.4	4.880	0.148	0.284	0.099	−83.9	g
propane	C_3H_8	44.10	369.8	4.248	0.200	0.276	0.152	−104.7	g
n-butane	C_4H_{10}	58.12	425.2	3.797	0.255	0.274	0.199	−125.7	g
iso-butane	C_4H_{10}	58.12	408.1	3.648	0.263	0.282	0.177	−134.2	g
n-pentane	C_5H_{12}	72.15	469.7	3.369	0.312	0.269	0.249	−146.7	g
iso-pentane	C_5H_{12}	72.15	460.4	3.381	0.306	0.270	0.228	−153.0	g
neo-pentane	C_5H_{12}	72.15	433.8	3.199	0.304	0.269	0.196	−168.1	g
n-hexane	C_6H_{14}	86.18	507.4	3.012	0.370	0.264	0.305	−166.9	g
n-heptane	C_7H_{16}	100.20	540.3	2.736	0.432	0.263	0.351	−187.7	g
n-octane	C_8H_{18}	114.23	568.8	2.486	0.492	0.259	0.396	−208.8	g
n-nonane	C_9H_{20}	128.26	595.7	2.306	0.548	0.255	0.438	−228.9	g
n-decane	$C_{10}H_{22}$	142.29	618.5	2.123	0.603	0.249	0.484	−249.5	g
cyclopropane	C_3H_6	42.08	397.9	5.575	0.163	0.274	0.134	53.4	g
cyclobutane	C_4H_8	56.11	459.9	4.985	0.210	0.274	0.187	27.2	g
cyclopentane	C_5H_{10}	70.13	511.8	4.502	0.258	0.273	0.194	−77.0	g
cyclohexane	C_6H_{12}	84.16	553.5	4.075	0.308	0.273	0.212	−123.1	g
alkenes									
ethene	C_2H_4	28.05	282.4	5.032	0.129	0.277	0.085	52.3	g
propene	C_3H_6	42.08	364.8	4.613	0.181	0.275	0.142	19.7	g
1-butene	C_4H_8	56.11	419.6	4.020	0.240	0.276	0.187	−0.5	g
1-pentene	C_5H_{10}	70.13	464.8	3.529	0.296	0.270	0.233	−20.9	g
1-hexene	C_6H_{12}	84.16	504.0	3.140	0.354	0.265	0.280	−41.7	g
1-heptene	C_7H_{14}	98.19	537.3	2.830	0.413	0.262	0.331	−62.3	g
1-octene	C_8H_{16}	113.32	566.6	2.550	0.472	0.256	0.375	−82.9	g
propadiene	C_3H_4	40.07	393.2	5.470	0.162	0.271	0.160	192.1	g
1,2-butadiene	C_4H_6	54.09	444.0	4.500	0.219	0.267	0.251	162.2	g
1,3-butadiene	C_4H_6	54.09	425.4	4.330	0.221	0.270	0.193	110.2	g
alkynes									
ethyne	C_2H_2	26.04	308.3	6.139	0.113	0.271	0.187	226.8	g
propyne	C_3H_4	40.07	402.4	5.628	0.164	0.276	0.216	185.4	g
1-butyne	C_4H_6	54.09	443.2	4.950	0.222	0.298	0.247	165.2	g
2-butyne	C_4H_6	54.09	488.2	5.080	0.221	0.277	0.130	146.3	g
1-pentyne	C_5H_8	68.12	481.2	4.170	0.277	0.289	0.290	144.4	g
1-hexyne	C_6H_{10}	82.15	516.2	3.620	0.322	0.272	0.333	123.7	g
aromatic compounds									
benzene	C_6H_6	78.11	562.2	4.898	0.259	0.271	0.211	82.9	g
toluene	$C_6H_5CH_3$	92.14	591.8	4.109	0.316	0.264	0.264	50.0	g
o-xylene	$C_6H_4(CH_3)_2$	106.17	630.4	3.734	0.369	0.263	0.313	19.0	g
m-xylene	$C_6H_4(CH_3)_2$	106.17	617.1	3.541	0.376	0.259	0.326	17.2	g
p-xylene	$C_6H_4(CH_3)_2$	106.17	616.3	3.511	0.379	0.260	0.326	18.0	g
ethylbenzene	$C_6H_5C_2H_5$	106.17	617.2	3.609	0.374	0.263	0.304	29.8	g
n-propylbenzene	$C_6H_5C_3H_7$	120.19	638.4	3.200	0.440	0.265	0.346	7.8	g
naphthalene	$C_{10}H_8$	127.17	748.4	4.051	0.413	0.269	0.302	150.6	g
anthracene	$C_{14}H_{10}$	178.23	873.0	2.900	0.554	0.221	0.489	230.1	g
phenanthrene	$C_{14}H_{10}$	178.23	869.3	2.900	0.554	0.222	0.495	207.1	g

Critical point and standard heat of formation data (cont.)

Compound		MW	T_C (K)	P_C (MPa)	$\hat{V}_C$ (m³ kmol⁻¹)	Z_C	ω	$\Delta \hat{h}_f^{\circ}$ (kJ mol⁻¹)	State
alcohols									
methanol	CH₃OH	32.04	512.6	8.096	0.118	0.224	0.566	− 200.7	g
ethanol	C₂H₅OH	46.07	516.3	6.384	0.167	0.248	0.637	− 234.4	g
1,2-ethanediol	CH₄(OH)₂	62.07	645.0	7.530	0.191	0.268	1.137	− 389.3	g
1-propanol	C₃H₇OH	60.10	536.7	5.170	0.219	0.253	0.628	− 256.4	g
2-propanol	C₃H₇OH	60.10	508.3	4.764	0.220	0.248	0.669	− 272.4	g
1-butanol	C₄H₉OH	74.12	562.9	4.413	0.275	0.259	0.595	− 274.7	g
1-pentanol	C₅H₁₁OH	88.15	586.2	3.880	0.326	0.260	0.594	− 298.7	g
1-hexanol	C₆H₁₃OH	102.18	611.4	3.510	0.381	0.263	0.580	− 317.6	g
1-heptanol	C₇H₁₅OH	116.20	631.9	3.150	0.435	0.261	0.587	− 341.6	g
1-octanol	C₈H₁₇OH	130.23	652.5	2.860	0.490	0.258	0.594	− 359.8	g
cyclohexanol	C₆H₁₁OH	100.16	625.2	3.749	0.322	0.232	0.514	− 294.6	g
phenol	C₆H₅OH	94.11	694.3	6.130	0.229	0.243	0.426	− 96.4	g
aldehydes									
formaldehyde	HCHO	30.03	408.0	6.586	0.105	0.204	0.282	− 115.9	g
acetaldehyde	CH₃CHO	44.05	461.0	5.550	0.157	0.227	0.317	− 166.2	g
propanal	C₂H₅CHO	58.08	496.0	4.660	0.210	0.237	0.302	− 186.0	g
butanal	C₃H₇CHO	72.11	525.0	4.000	0.263	0.241	0.345	− 206.1	g
benzaldehyde	C₆H₅CHO	106.12	695.0	4.650	0.324	0.261	0.305	− 36.8	g
ketones									
acetone	CH₃COCH₃	58.08	508.2	4.702	0.209	0.233	0.306	− 217.2	g
methyl ethyl ketone	C₂H₅COCH₃	72.11	535.5	4.154	0.267	0.249	0.324	− 238.4	g
methyl n-propyl ketone	C₃H₇COCH₃	86.13	561.1	3.694	0.301	0.238	0.346	− 259.2	g
ethers									
dimethyl ether	CH₃OCH₃	46.07	400.1	5.370	0.170	0.274	0.200	− 184.1	g
methyl ethyl ether	CH₃OC₂H₅	60.10	437.8	4.398	0.221	0.267	0.219	− 216.4	g
diethyl ether	C₂H₅OC₂H₅	74.12	466.7	3.638	0.280	0.262	0.285	− 252.1	g
carboxylic acids									
formic acid	HCOOH	46.03	580.0	7.390	0.125	0.192	0.473	− 378.6	g
acetic acid	CH₃COOH	60.05	592.7	5.786	0.171	0.201	0.462	− 432.3	g
propanoic acid	C₂H₅COOH	74.08	612.0	5.370	0.230	0.243	0.544	− 453.5	g
butanoic acid	C₃H₇COOH	88.11	628.0	4.420	0.283	0.240	0.604	− 470.3	g
pentanoic acid	C₄H₉COOH	102.13	651.0	3.810	0.336	0.237	0.627	− 490.4	g
benzoic acid	C₆H₅COOH	122.12	751.0	4.470	0.344	0.246	0.604	− 290.2	g
oxalic acid	COOHCOOH	90.04	804.0	7.020	0.205	0.215	0.918	− 723.7	g
esters									
methyl formate	HCOOCH₃	60.05	487.2	5.998	0.172	0.255	0.254	− 352.4	g
ethyl methanoate	HCOOC₂H₅	74.08	508.4	4.742	0.229	0.257	0.285	− 388.3	g
methyl acetate	CH₃COOCH₃	74.08	506.8	4.690	0.228	0.254	0.325	− 410.0	g
ethyl ethanoate	CH₃COOC₂H₅	88.11	523.3	3.830	0.286	0.252	0.361	− 442.9	g
propyl ethanoate	CH₃COOC₃H₇	102.13	549.4	3.360	0.345	0.254	0.394	− 464.8	g
nitro compounds									
nitromethane	CH₃NO₂	61.04	588.2	6.313	0.173	0.224	0.348	− 74.7	g
nitroethane	C₂H₅NO₂	75.07	593.0	5.160	0.236	0.247	0.368	− 101.3	g
1-nitropropane	C₃H₇NO₂	89.09	605.0	4.350	0.288	0.249	0.412	− 124.7	g
nitrobenzene	C₆H₅NO₂	123.11	719.0	4.400	0.349	0.257	0.448	67.6	g

Critical point and standard heat of formation data (cont.)

Compound		MW	T_C (K)	P_C (MPa)	$\hat{V}_C$ (m³ kmol⁻¹)	Z_C	ω	$\Delta \hat{h}_f^0$ (kJ mol⁻¹)	State
halides									
bromomethane	CH_3Br	94.94	467.0	8.000	0.156	0.321	0.192	−37.7	g
chloromethane	CH_3Cl	50.49	416.3	6.679	0.139	0.268	0.153	−82.0	g
di-chloromethane	CH_2Cl_2	84.93	510.0	6.080	0.185	0.265	0.192	−95.5	g
tri-chloromethane	$CHCl_3$	119.38	536.4	5.472	0.239	0.293	0.213	−103.2	g
tetra-chloromethane	CCl_4	153.82	556.4	4.560	0.276	0.272	0.193	−96.0	g
fluoromethane	CH_3F	34.03	317.7	5.877	0.113	0.251	0.204	−237.7	g
di-fluoromethane	CH_2F_2	52.02	351.6	5.830	0.121	0.241	0.276	−452.7	g
tri-fluoromethane	CHF_3	70.01	298.9	4.836	0.133	0.259	0.267	−693.3	g
tetra-fluoromethane	CF_4	88.01	227.5	3.739	0.140	0.277	0.186	−933.2	g
bromoethane	C_2H_5Br	108.97	503.8	6.232	0.215	0.320	0.253	−63.6	g
chloroethane	C_2H_5Cl	64.51	460.4	5.269	0.200	0.275	0.191	−112.3	g
mercaptans									
methyl mercaptan	CH_3SH	48.11	467.0	7.235	0.145	0.268	0.146	−22.3	g
ethyl mercaptan	C_2H_5SH	62.14	499.2	5.490	0.207	0.274	0.192	−45.8	g
n-propyl mercaptan	C_3H_7SH	76.16	536.0	4.630	0.254	0.264	0.235	−67.5	g
iso-propyl mercaptan	C_3H_7SH	76.16	517.0	4.750	0.254	0.281	0.212	−75.9	g
n-butyl mercaptan	C_4H_9SH	90.19	569.0	3.970	0.307	0.258	0.278	−88.1	g
n-pentyl mercaptan	$C_5H_{11}SH$	104.22	598.0	3.470	0.359	0.251	0.321	−109.8	g
inorganic compounds									
argon	Ar	39.95	150.9	4.898	0.075	0.291	0.000	−143.9	g
bromine	Br_2	159.81	584.2	10.335	0.135	0.287	0.119	30.9	g
carbon monoxide	CO	28.01	132.9	3.499	0.093	0.295	0.066	−110.5	g
carbon dioxide	CO_2	44.01	304.2	7.382	0.094	0.274	0.228	−393.5	g
chlorine	Cl_2	70.91	417.2	7.711	0.124	0.275	0.069	0.0	g
hydrochloric acid	HCl	36.46	324.7	8.309	0.081	0.249	0.132	−92.3	g
fluorine	F_2	38.00	144.3	5.215	0.066	0.288	0.059	0.0	g
hydrofluoric acid	HF	20.01	461.2	6.485	0.069	0.117	0.383	−272.6	g
helium-4	He	4.00	5.2	0.228	0.057	0.302	−0.390	0.0	g
hydrogen	H2	2.02	33.2	1.313	0.064	0.305	−0.215	0.0	g
water	H_2O	18.02	647.1	22.055	0.056	0.229	0.345	−241.8	g
hydrogen peroxide	H_2O_2	34.02	730.2	21.684	0.077	0.278	0.360	−136.1	g
iodine	I_2	253.81	819.2	11.654	0.155	0.265	0.117	62.4	g
hydrogen iodide	HI	127.91	423.9	8.310	0.122	0.288	0.038	26.4	g
krypton	Kr	83.80	209.4	5.502	0.091	0.288	0.000	0.0	g
neon	Ne	20.18	44.4	2.653	0.042	0.300	−0.414	0.0	g
nitrogen	N_2	28.01	126.1	3.394	0.090	0.292	0.040	0.0	g
nitric oxide	NO	30.01	180.2	6.485	0.058	0.250	0.585	90.2	g
nitrogen dioxide	NO_2	46.01	431.4	10.133	0.082	0.233	0.849	33.1	g
nitrous oxide	N_2O	44.01	309.6	7.245	0.097	0.274	0.142	82.0	g
ammonia	NH_3	17.03	405.7	11.278	0.072	0.242	0.252	−45.9	g
nitric acid	HNO_3	63.01	520.0	6.890	0.145	0.231	0.714	−134.3	g
oxygen	O_2	32.00	154.6	5.043	0.073	0.288	0.022	0.0	g
ozone	O_3	48.00	261.0	5.573	0.089	0.229	0.227	142.7	g
sulphur	S	32.02	1313.0	18.208	0.158	0.264	0.262	278.8	g
hydrogen sulphide	H_2S	34.08	373.5	8.963	0.098	0.284	0.083	−20.6	g
sulphur oxide	SO_2	64.07	430.8	7.884	0.122	0.269	0.245	−296.8	g

Other important data

Air properties

Composition of clean, dry air :

78.0849 mol% N_2
20.9479 mol% O_2
0.9340 mol% Ar
0.0314 mol% CO_2
0.0018 mol% Ne

Average molecular weight :	28.965
Pseudo-critical temperature :	132.4 K
Pseudo-critical pressure :	3.969 MPa
Pseudo-critical specific volume :	0.0795 m^3 $kmol^{-1}$
Acentric factor :	0.039

Martian atmosphere properties

Composition of dry atmosphere :

95.49 mol% CO_2
2.70 mol% N_2
1.60 mol% Ar
0.13 mol% O_2
0.08 mol% CO

Average molecular weight :	43.485
Pseudo-critical temperature :	296.61 K
Pseudo-critical pressure :	7.243 MPa
Pseudo-critical specific volume :	0.0923 m^3 $kmol^{-1}$
Acentric factor :	0.216

Avogadro constant	$6.022\ 1 \times 10^{23}\ mol^{-1}$
Boltzmann constant	$1.380\ 7 \times 10^{-23}\ J\ K^{-1}$
Faraday constant	$9.648\ 5 \times 10^{4}\ C\ mol^{-1}$
Stefan-Boltzmann constant	$5.670\ 5 \times 10^{8}\ Wm^{-2}\ K^{-4}$
Universal gas constant	$8.314\ 5\ J\ mol^{-1}\ K^{-1}$

Nomenclature

C_P	specific heat capacity at constant pressure, kJ kg^{-1} K^{-1}
C_{PL}	specific heat capacity of saturated liquid, kJ kg^{-1} K^{-1}
C_{PV}	specific heat capacity of saturated vapour, kJ kg^{-1} K^{-1}
C_V	specific heat capacity at constant volume, kJ kg^{-1} K^{-1}
h	specific enthalpy, kJ kg^{-1}
h_E	specific enthalpy of ethane, kJ kg^{-1}
h_f	specific enthalpy of saturated liquid, kJ kg^{-1}
h_{fg}	latent heat of vaporization, kJ kg^{-1}
h_g	specific enthalpy of saturated vapour, kJ kg^{-1}
h_M	specific enthalpy of methane, kJ kg^{-1}
h_P	specific enthalpy of propane, kJ kg^{-1}
$\Delta \hat{h}_f$	standard heat of formation, kJ mol^{-1}
k	thermal conductivity, mW m^{-1} K^{-1}
k_L	thermal conductivity of saturated liquid, mW m^{-1} K^{-1}
k_V	thermal conductivity of saturated vapour, mW m^{-1} K^{-1}
p	vapour pressure, kPa, MPa
P	pressure, kPa, MPa
P_C	critical pressure, MPa
P_V	vapour pressure, kPa, MPa
R	Universal gas constant, J mol^{-1} K^{-1}
S	specific entropy, kJ kg^{-1} K^{-1}
S_f	specific entropy of saturated liquid, kJ kg^{-1} K^{-1}
S_g	specific entropy of saturated vapour, kJ kg^{-1} K^{-1}
S_{fg}	difference in saturated liquid and vapour specific entropies, kJ kg^{-1} K^{-1}
t	temperature, °C
T	absolute temperature, K
T_C	critical temperature, K
T_R	reduced temperature
$\hat{V}_C$	critical specific volume, m^3 kmol^{-1}
Z	compressibility factor
Z_c	critical compressibility factor
γ	ratio of specific heat capacities
μ	viscosity, μPa
μ_L	viscosity of saturated liquid, μPa
μ_V	viscosity of saturated vapour, μPa
v_g	gas specific volume, m^3 kg^{-1}
ρ	density, kg m^{-3}
ρ_L	density of saturated liquid, kg m^{-3}
ρ_V	density of saturated vapour, kg m^{-3}
ω	acentric factor